Progress in Optical Science and Photonics

Volume 16

The purpose of the series Progress in Optical Science and Photonics is to provide a forum to disseminate the latest research findings in various areas of Optics and its applications. The intended audience are physicists, electrical and electronic engineers, applied mathematicians, biomedical engineers, and advanced graduate students.

More information about this series at http://www.springer.com/series/10091

Swapna S. Nair · Reji Philip

Nanomaterials for Luminescent Devices, Sensors, and Bio-imaging Applications

 Springer

Swapna S. Nair
Central University of Kerala
Periye, Kerala, India

Reji Philip
Raman Research Institute
Bengaluru, Karnataka, India

ISSN 2363-5096   ISSN 2363-510X (electronic)
Progress in Optical Science and Photonics
ISBN 978-981-16-5369-8   ISBN 978-981-16-5367-4 (eBook)
https://doi.org/10.1007/978-981-16-5367-4

This Springer imprint is published by the registered company Springer Nature Singapore Pte Ltd.
The registered company address is: 152 Beach Road, #21-01/04 Gateway East, Singapore 189721,
Singapore

Preface

Nanomaterials and nanotechnology will permeate all aspects of everyday life in the coming decades. Right from a multitude of memory and storage devices, displays, sensors and transducers, its impact will be phenomenal in the making of nanoscale energy harvesters and tiny implantable devices consisting of nanomotors, nanogears, nanosensors, etc., which can be installed inside the human body while being invisible to naked eyes and even to light microscopes. Exploration of novel nanomaterials with multifunctional properties essential for technological applications and fabrication of functional devices based on them will assume major share in the semiconductor electronics market.

This era is of quantum dots (QDs) and QD-based optical devices including LASERs and LEDs. The tuneable optical properties of these quantum systems make them ideal candidates for designing optical biosensors and bioimaging probes as well. However, their enormous production cost is a great challenge for the popularisation of the technology, and therefore, development of novel wet chemistry strategies for the fabrication of ultrafine nanosystems such as quantum dots, quantum cubes and quantum cages is the need of the hour.

Chapter 2 of this book provides a general review on the optical properties of nanomaterials, especially the quantum confined systems, and explains the scope of such materials in designing and fabrication of different devices like QD LEDs, QDs-based display devices, LASERS, optical sensors and finally will explain their potential in in vitro and in vivo bioimaging.

Tuneable optical band gap is the most appealing property of semiconductor nanoparticles. Especially when the nanoparticles are in the strong confinement regime, (1–5 nm range for most of the semiconductors), tailored band gap can be induced and band engineered nanosystems with core shell geometry can be exploited for their vast technological applications in designing QD LEDs, QD LASERS and display devices by incorporation of defect states that falls inside the band gap, which can create systems with tailor made emissions. Size, surface functionalization, shape, the level of doping and the dielectric environments are the deciding parameters which needs thorough tuning and monitoring. Chapter 2 focuses on the applications of semiconductor quantum structures for luminescent devices like LEDs and LASERs.

Metal nanoparticles possess excellent application potential in optical devices due to their surface plasmon resonance (SPR) and localised surface plasmon resonance (LSPR) properties. Noble metal nanoparticles like gold and silver are the regular choice for these. Plasmonic copper NPs emerge as an economic alternative to noble metals like gold and silver. The metal nanoparticles with enhanced LSPR properties can offer extensive direct application in surface enhanced Raman scattering and related applications like ultra-sensitive detection of molecules. Chapter 3 focuses on the plasmonic nanosystems and their applications in SERS.

Apart from the technological applications of nanomaterials in sensing and luminescent devices, their application in cellular and live bioimaging is also commendable. Search for novel materials and geometries with low toxicity and high luminescence are underway, and non-conventional materials like fluorescent carbon, metallic systems other than gold and silver, core shell semiconductor systems, dye tagged nanosystems, etc. are being developed and employed for biosensing and imaging applications. Chapter 4 discusses the applications of nanomaterials in biology and medicine, especially in biosensing and imaging.

Chapter 6 discusses the nonlinear optical properties of nanomaterials. Quite often, nanomaterials are found to show enhanced optical nonlinearities compared to their bulk counterparts. Metal nanoparticles are known to exhibit both saturable absorption and reverse saturable absorption depending on the wavelength of excitation, optical intensity and sample concentration. While saturable absorbers are essential for sub-nanosecond laser pulse generation, reverse saturable absorbers are potentially useful for protecting human eyes and sensitive detectors from hazardous levels of laser radiation. Semiconductor quantum dots also exhibit interesting nonlinear optical behaviour. The electronic origin of optical nonlinearities and some of the experimental techniques for measuring the nonlinearities are discussed in this chapter.

This book is a comprehensive version of an interdisciplinary research approach towards nanomaterials and its applications, which is the need of the hour. The authors are experts in the area of synthesis, optical characterization and biological application of metal and semiconductor nanoparticles. There exists a knowledge gap between the nanomaterial synthesis and its applications, especially in biological systems ranging from cellular models to whole living systems. While designing the applications, the safety of the environment needs to be ensured by the customisation of nanomaterials for grounding the toxicity to zero percentage. This book is aimed to bridge the gap apart from imparting the technical knowledge in the field to the readers. The authors wish that the book will serve as a knowledge platform for the researchers in the field of nanomaterials and its applications, whereby innovative thoughts will be stimulated to achieve the major ongoing goals of the century such as Internet of things (IoT).

Periye, India   Swapna S. Nair
Bengaluru, India   Reji Philip

Contents

Chapter 1
Introduction to the Optical Applications of Nanomaterials

Swapna S. Nair

1.1 Introduction

From time immemorial, optical properties of materials were fascinating for the mankind. Colored materials were identified and were used for paints while transparent materials were widely sought after for making decorative lights and light shades and windows. Reflecting surfaces of metals were widely exploited for mirror applications. Silvered mirrors were later invented by semi/full silvering of transparent materials for partial /full reflectance mirrors wherever it was needed. Several thousands of years later, luminous materials were discovered and were remained as a point of wonder for many centuries. Materials were developed and improved and playing with their optical properties like absorption, luminescence, reflection, dispersion and scattering effects beautiful glass-based display items were generated even during ancient times. Hence this chapter summarizes overall optical properties and related applications of nanomaterials and will introduce the promising materials for each of those application.

1.1.1 Optical Absorption, Scattering and Emission in Nanomaterials

Optical absorption and emission are the most important optical properties of materials which is strongly dependent on the band structure and band gap. These are important parameters not only from the fundamental point of view but also from the device application point of view particularly in deciding their potential in promising

S. S. Nair (✉)
Department of Physics, Central University of Kerala, Kasaragod 671320, India

optoelectronic applications especially LED, LASERS, photo detectors and photo-voltaic devices [1–4]. Nanomaterials, especially quantum structures like Quantum dots (QDs), quantum wires (QW), quantum wells (QWl) etc. offer superior potential as far as the optoelectronic applications are concerned due to their tunable optical properties when compared to their bulk cousins. Their optical properties can be tailored by varying grain size [5], surface functionalization [6], shape and core–shell type formulations as well as their ultrafine size helps us in device miniaturization, which enhances their application potential by several folds.

About the extinction, for ultrafine particles, the extinction is almost fully due to absorption and contribution from scattering is too negligible. When the particle size is above several tens of nanometers, the contribution of scattering also becomes as a major one and above 100 nm size range, the dominant contribution is from scattering and hence by tuning the particle size, it is possible for us to tune the scattering and absorption contributions [7]. Aggregation is another phenomenon which induces scattering and hence charge stabilized or surfactant coated QDs dispersed in a base fluid can greatly reduce the scattering contribution and hence often their colloidal suspensions look transparent while the aggregated quantum dots exhibit turbid looks.

Emissions including Fluorescence and phosphorescence are the other major exciting properties of materials. In nanomaterials especially in quantum structures, tailorable emission and strong quantum yield for fluorescence is often observed which makes them efficient candidates for luminescent devices like LEDs and LASERs. Nowadays research is in progress in fabricating all solution processed inorganic QD based LED displays which can drastically reduce the production cost as well as can ensure better stability and long life for the displays compared to their organic LED counter parts [8].

1.1.1.1 Semiconductor Quantum Dots

At ultrafine particle size regime, the size and shape of the particle plays major role in predicting the optical absorption and fluorescence in nanoparticles and the band gap will be greatly blue shifted from the bulk value due to the confinement energy and its possible to have a spectrum of QDs with different size emitting in all possible wavelength regime from IR to UV with proper tailoring of size shape and surface functionalities. Cadmium based semiconductor quantum dots especially CdS and CdSe often exhibit high degree of emission wavelength tailorability by tuning the particle size and introducing the core shell structure [9]. Figure 1.1 shows the CdS nanoparticles with particle size varying from 5 ± 0.5 nm to 20 ± 2 nm, which was synthesized in our lab. Shape also is another factor which can modify the optical properties in semiconductor quantum structures [10]. The emission wavelength can be shifted across the visible spectrum, with the smaller particles emitting in the blue and the larger particles emitting red light.

Organic dyes-based materials were obvious choices for fluorescence and related applications in the pre quantum structure era due to their strong absorption and emission in the visible range and existence of a large spectrum of combination dyes

Fig. 1.1 CdS nanoparticles with particle size varying from 5 ± 0.5 to 20 ± 2 nm

for effective tailoring of the emission [11]. However, since their invention, QDs and other quantum structures started to play a dominant share in the luminescent device market due to their excellent properties like high photo bleach threshold, large quantum yileds and much wider excitation bandwidth (which result in a large stokes shift). Apart from these, higher stability and shelf life are other major advantages which makes them excellent for their applications in imaging and displays. Toxicity is a major factor which needs to be addressed before employing these semiconductor quantum dots for bio imaging applications and novel biocompatible quantum dots are being developed for this purpose and methods are being developed to reduce the toxicity of the already existing materials too [12].

QD, QW and Qwl based LEDs and Lasers are already in market and employment of them in photovoltaic and photo detector devices are also reported extensively [13]. Their colour as well as fluorescence emission wavelength tunability makes them excellent for luminescent devices as well as photovoltaics. CdS QDs with different particle size can show different color, which was synthesized and confirmed in our lab (Fig. 1.2).

New materials and their effective combinations are being sought after for making them fit for specific applications and demand for novel nanomaterials with high stability, tunable emission under wider excitation band width and lower toxicity is the challenge of the hour. Figure 1.3 shows the band edge tuning in CdS NPs by varying synthesis parameter.

Fig. 1.2 CdSe quantum dots in varying particle sizes (ranging from 2 ± 0.45 nm to 7.5 ± 0.5 nm)

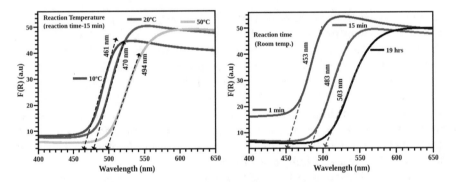

Fig. 1.3 Band edge tuning in CdS NPs by varying synthesis parameters

1.1.2 Metal Nanoparticles and Surface Plasmon Resonance (SPR)

Metallic nanoparticles exhibit often exciting optical properties and have been a research focus in the last decade especially due to their surface plasmon resonance (SPR) [14]. Noble metal nanoparticles like gold, silver and platinum are widely researched for their plasmonic properties and recently copper also emerged as an economic alternative to the noble metal nanoparticles [15]. When the frequency of collective oscillations of the conduction band electrons of the metal become equal to the applied frequency (incident photon frequency), resonance will occur and the particle will strongly absorb or scatter energy, which is termed as surface plasmon resonance [14]. Mostly, noble metal nanoparticles like gold and silver can support SPR modes and hence exhibit strong absorption in the visible range. Particle size, core- shell structure, surface charges, dielectric properties of the medium, aggregation effects etc. strongly influence the SPR in metallic nanoparticles [16]. This property results in visibly bright color in noble metal NPs whose wavelength is tunable by suitably playing with the particle size, surface charges etc. Multi core-shell formation is another brilliant attempt made by researchers which could yield bright coloured metallic NPs with excellent tailorable colour gradient (Fig. 1.4).

Recently metal–metal oxide core shells are fabricated to obtain SPR in the near IR range for NIR active devices which can offer wide applications in tumor therapy [18]. Localized surface plasmon resonance is another promising research area where the high surface area of small metallic nanoparticles can be exploited (Fig. 1.5).

Fig. 1.4 Fluorescence under UV illumination in Cu-GSH NCs through aggregate induced emission [17]

Fig. 1.5 Copper 2D plates and spirals with excellent optical and electrical properties. Authors unpublished work

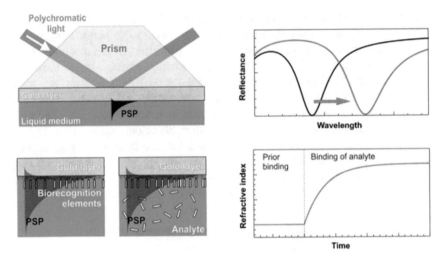

Fig. 1.6 Concept of surface plasmon resonance biosensor. Reprinted with permission from Ref [22] Copyright (2013) Elsevier

1.1.2.1 Surface Enhanced Raman Scattering

Surface Enhanced Raman Scattering (SERS) is a surface sensitive technique used for the detection of molecules by enhancement of Raman scattering signal by the molecules adsorbed on the SERS active surfaces like metallic gold silver nanoparticles etc. [19, 20]. In Raman scattering, the scattered photons contain information about the vibrational modes of the molecule which is used for investigation. The finger print region of the Raman spectrum can provide vital information about the materials including the vibrational modes in the molecules [21]. However, efficiency is often very low due to the poor cross section as compared to the Rayleigh cross section. However, if the molecules are attached to SPR active substrates, the cross section of Raman scattering is much higher than the cross sections of Rayleigh scattering/fluorescence and hence efficient detection of molecules can be made possible [21] (Fig. 1.6).

1.1.3 Semiconductor Nanoparticles for Display Devices

Luminescent devices is often fabricated with cadmium based nanomaterials, because of their bright and tunable luminescence in the visible range. Additionally they exhibit high external quantum efficiency (EQE) and lifetime. But other materials are essential to cover the blue region of the spectrum. Toxicity is another major issue which place them in the back seat and novel core shell alternatives are being developed in which ZnS and ZnSe are emerging as hot candidates. High life time and tunable wide

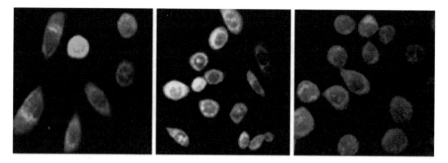

Fig. 1.7 Fluorescent images of chines hamster overy with CdTe/CdS/ZnS QD [25]

spectrum emission including the blue and near UV region of the spectrum makes them next generation material for QD LED displays. Reported EQE are n the range of 12–21%. QDs are also ideal for photovoltaic devices due to their tailorable band edge and high conversion efficiency. Combinations of organic and QD layers is reported to provide very high efficiency.

1.1.4 Semiconductor and Metallic Nanoparticles for Bioimaging

Gold NPs were the most widely researched material for both in vivo and in vitro bio imaging due to the low levels of toxicity of gold NPs [23]. However, more economic alternatives including copper is being investigated in these directions recently. Semiconductor NPs like CdS, CdSe etc. are not favored for bioimaging applications due to their high toxicity levels of cadmium compounds. ZnO and ZnS are futuristic materials in this angle, which alone or as core shell parts of cadmium-based compounds could greatly reduce the toxicity and can offer bright promise for next generation bio medical applications [24, 25]. NPs are often easily up-taken by the cells through endocytosis and hence can be exploited for their application as tag-less imaging probes. High photo bleach threshold is another major advantage of NPs which makes them ideal candidates compared to the conventional dye-based imaging [26]. NPs surfaces can be functionalized so as to have targeted cellular imaging and selective organelle marking also can be done using NPs [27] (Fig. 1.7).

1.1.5 Non Linear Optical Properties of Materials

Nanomaterials are reported to exhibit enhanced optical nonlinearities compared to their bulk counterparts. Metal nanoparticles exhibit both saturable absorption and reverse saturable absorption which is strongly dependent on the excitation wavelength, sample concentration, aggregation, particle size and optical intensity. Both saturable absorbers and reverse saturable absorbers find extensive applications in devices. The former offers potential for sub-nanosecond laser pulse generation, while the latter is effective in fabrication of devices for protecting human eyes and sensitive detectors from high intense laser radiation. Semiconductor quantum dots like fullerenes [28, 29], and ferrofluids also exhibit excellent non-linear properties [30, 31].

This book aims at providing detailed updates on the research activities that are being performed across the world on the optical applications of semiconductor and metallic nanoparticles. Thrust is provided to their applications in SPR, SERS, luminescent devices and bio imaging. Non linear optical applications of NPs are also elaborated. The following are the major and sub themes of the book.

1.2   Conclusion

Novel materials are being developed to replace toxic as well as costly ones in variety of optical applications including photovoltaics, LEDs, SPR based sensors and SERS substrates. Materials of obvious choice like gold and silver are being replaced by copper and other core shell systems in SPR and SERS applications while cadmium as well as lead based materials are under less demand for the semiconductor industry. Search for novel materials and their combinations are the need of the hour. Regarding the display devices, low shelf life and stability being the obvious negative points, organic LEDs and PV devices are also under threat and their quantum dot counterparts are going to rule the market soon. However, bringing down of their cost is a big issue.

References

1. V. Wood, V. Bulović, Colloidal quantum dot light-emitting devices. Nano. Rev. **1**, 5202 (2010)
2. P.S. Zory, Optical gain in III-V bulk and quantum well semiconductors, in *Quantum Well Lasers* (Academic, 1993), pp. 17–96
3. J.C. Cao, H.C. Liu, Terahertz semiconductor quantum well photodetectors, in *Semiconductors and Semimetals*, vol. 84, (Elsevier, 2011), pp. 195–242
4. I. Sayed, S.M. Bedair, Quantum well solar cells: principles, recent progress, and potential. IEEE J. Photovoltaics **9**, 402–423 (2019)
5. K.-F. Lin, H.-M. Cheng, H.-C. Hsu, L.-J. Lin, W.-F. Hsieh, Band gap variation of size-controlled ZnO quantum dots synthesized by sol–gel method. Chem. Phys. Lett. **409**, 208–211 (2005)
6. O.M. Bankole, O.J. Achadu, T. Nyokong, Nonlinear interactions of zinc phthalocyanine-graphene quantum dots nanocomposites: investigation of effects of surface functionalization with heteroatoms. J. Fluoresc. **27**, 755–766 (2017)
7. J.X. Xu et al., Quantification of the photon absorption, scattering, and on-resonance emission properties of CdSe/CdS core/shell quantum dots: Effect of shell geometry and volumes. Anal. Chem. **92**, 5346–5353 (2020)

8. Y.-M. Huang et al., Advances in quantum-dot-based displays. Nanomaterials **10**, 1327 (2020)
9. H. Han, G.D. Francesco, M.M. Maye, Size control and photophysical properties of quantum dots prepared via a novel tunable hydrothermal route. J. Phys. Chem. C **114**, 19270–19277 (2010)
10. D. Vanmaekelbergh et al., Shape-dependent multiexciton emission and whispering gallery modes in supraparticles of CdSe/multishell quantum dots. ACS Nano **9**, 3942–3950 (2015)
11. U. Resch-Genger, M. Grabolle, S. Cavaliere-Jaricot, R. Nitschke, T. Nann, Quantum dots versus organic dyes as fluorescent labels. Nat. Methods **5**, 763 (2008)
12. M. Bayal et al., Cytotoxicity of nanoparticles-Are the size and shape only matters? or the media parameters too?: a study on band engineered ZnS nanoparticles and calculations based on equivolume stress model. Nanotoxicology **13**, 1005–1020 (2019)
13. J. Wu, S. Chen, A. Seeds, H. Liu, Quantum dot optoelectronic devices: lasers, photodetectors and solar cells. J. Phys. D. Appl. Phys. **48**, 363001 (2015)
14. T.A. El-Brolossy et al., Shape and size dependence of the surface plasmon resonance of gold nanoparticles studied by Photoacoustic technique. Eur Phys J: Spec Top **153**, 361–364 (2008)
15. D. Mott, J. Galkowski, L. Wang, J. Luo, C.J. Zhong, Synthesis of size-controlled and shaped copper nanoparticles. Langmuir **23**, 5740–5745 (2007)
16. V. Amendola, R. Pilot, M. Frasconi, O.M. Maragò, M.A. Iatì, Surface plasmon resonance in gold nanoparticles: A review. J. Phys. Condens. Matter **29**, 203002 (2017)
17. N. Chandran et al., Label free, nontoxic Cu-GSH NCs as a nanoplatform for cancer cell imaging and subcellular pH monitoring modulated by a specific inhibitor: bafilomycin A1. ACS Appl. Bio Mater. **3**, 1245–1257 (2020)
18. X. Huang, I.H. El-Sayed, W. Qian, M.A. El-Sayed, Cancer cell imaging and photothermal therapy in the near-infrared region by using gold nanorods. J. Am. Chem. Soc. **128**, 2115–2120 (2006)
19. Q. Yu et al, Surface-enhanced Raman scattering on gold quasi-3D nanostructure and 2D nanohole arrays. Nanotechnology **21**, 355301 (2010)
20. S.J. Oldenburg, S.L. Westcott, R.D. Averitt, N.J. Halas, Surface enhanced Raman scattering in the near infrared using metal nanoshell substrates. J. Chem. Phys. **111**, 4729–4735 (1999)
21. R. Pilot, R. Signorini, L. Fabris, Surface-enhanced Raman spectroscopy: Principles, substrates, and applications. *Metal Nanoparticles and Clusters: Advances in Synthesis, Properties and Applications* (2017). https://doi.org/10.1007/978-3-319-68053-8_4
22. H. Šípová, J. Homola, Surface plasmon resonance sensing of nucleic acids: a review. Anal. Chim. Acta **773**, 9–23 (2013)
23. C.J. Murphy et al., Gold nanoparticles in biology: beyond toxicity to cellular imaging. Acc. Chem. Res. **41**, 1721–1730 (2008)
24. Y. Li et al., Cellulosic micelles as nanocapsules of liposoluble CdSe/ZnS quantum dots for bioimaging. J. Mater. Chem. B **4**, 6454–6461 (2016)
25. C. Yan et al., Synthesis of aqueous CdTe/CdS/ZnS core/shell/shell quantum dots by a chemical aerosol flow method. Nanoscale Res. Lett. **5**, 189–194 (2010)
26. M. Bruchez, M. Moronne, P. Gin, S.Weiss, A.P. Alivisatos, Semiconductor nanocrystals as fluorescent biological labels. Science **281**, 2013–2016 (1998)
27. V. Bagalkot et al., Quantum dot-aptamer conjugates for synchronous cancer imaging, therapy, and sensing of drug delivery based on bi-fluorescence resonance energy transfer. Nano Lett. **7**, 3065–3070 (2007)
28. A.B. Bourlinos et al., Green and simple route toward boron doped carbon dots with significantly enhanced non-linear optical properties. Carbon N. Y. **83**, 173–179 (2015)
29. P. Innocenzi, B. Lebeau, Organic–inorganic hybrid materials for non-linear optics. J. Mater. Chem. **15**, 3821–3831 (2005)
30. S. Nair, J. Thomas, S. Sandeep, M.R. Anantharaman, R. Philip, Non linear optical properties of ferrofluids investigated by Z scan technique. Chaos Complex. Lett. **7**, 55 (2013)
31. J. Philip, J.M. Laskar, Optical properties and applications of ferrofluids—a review. J. nanofluids **1**, 3–20 (2012)

Chapter 2
Quantum Wells, Wires and Dotes for Luminescent Device Applications

Manikanta Bayal, Neeli Chandran, Rajendra Pilankatta, and Swapna S. Nair

2.1 Introduction

Luminescent nanomaterials are gaining much research thrust owing to their vast applications in optical devices while their quantum structures like Quantum dots (QDs), quantum wells, and quantum wires find extensive research potential due to their sharp and discrete like energy states resulting in enhanced monochromaticity and better quantum yield. The superior optical and electronic properties of organic and inorganic nanomaterials are the main reason behind the thought of device application. The type of electronic confinement decides the nomenclature of the nanostructures. Spherical NPs with particle size below few nms (depends on the Bohr radius) is said to be QDs [1, 2] and it has excellent optical properties including band gap tenability [3], high efficiency [4] and narrow emission band width [5–7]. ZnSe, ZnS, ZnO, CdS, CdSe, GaAs etc. are some classic examples for semiconductor QDs, which shows high applications in the area of optical devices [8, 9]. The most fascinating property of a QD is, the tunability of band gap from IR to even the UV range by efficient tailoring of grain size [10]. In nanoscale materials the confinement of individual electrons and holes dominate. Therefore, the optical and electrical properties of materials become size and shape dependent in this regime. In general, three regimes of quantization can be defined depending on whether the charge carriers are confined in 1, 2, or 3 dimensions. Confinement in 1-D creates structures that have been termed as quantum wells; this is because the first such structure could be described by the elemental quantum mechanics of a particle in a one-dimensional

M. Bayal · N. Chandran · S. S. Nair (✉)
Department of Physics, Central University of Kerala, Kasaragod 671320, India

R. Pilankatta
Department of Biochemistry and Molecular Biology, Central University of Kerala, Kasaragod 671320, India

box; these structures have also been labelled as "quantum films". Carrier confinement in 2-Ds produces "quantum wires", and in 3-D produces "quantum dots (QDs)" (also called nanocrystals or quantum boxes).

The dimensionality of confinement affects many aspects of quantization. Since their invention by Esaki and Tsu in the 1970s [11], semiconductor quantum wells and superlattices have evolved from scientific curiosities to a means of probing the fundamentals of quantum mechanics, and more recently into the fabrication of semiconductor devices which are the need of the hour. For an electron in vacuum away from the influence of electromagnetic fields, the total energy is just the kinetic energy.

It is always desirable to use the spherical potential well with infinite potential as the model system to deal with the quantum confinement in nanocrystals, while the electron and hole effective masses are considered to be isotropic. The extreme cases are the weak and strong confinement regimes while intermediate confinement regime comes in between both. In the weak confinement regime, the particle size is often below 10 nm, but still it is higher than 4 times the Bohr radius a_B, in which we can expect the quantisation of exciton centre of mass. In contrary to that, in the strong confinement regime, the grain size is much less than the a_B which results in a condition that there cannot be a hydrogen like exciton bound state for the electron and hole. It is also to be noted that the zero point kinetic energy (KE) of electron and hole due to strong confinement is so huge which will result in a large shift in optical band gap of the material. At this size range, the band gap tunability range is so big and even the materials with bulk band edge lying in the IR range can be brought to the blue or even UV range. But to realise this in reality, fabrication of QDs with monodisperse grains [12] is the required [12].

This chapter deals with the luminescent device applications of different nanostructures such as QDs, quantum wells, quantum wires etc. QDs can be synthesized using chemical methods like co-precipitation, sol–gel and hydrothermal method. Co-precipitation is a simple chemical synthesis technique, where anionic and cationic precursors are used to form QDs. ZnS QDs can be easily synthesise using this method by taking sodium sulphide and zinc chloride as anionic and cationic precursors respectively. Quantum wells and wires can be synthesised using template assisted method. Porus alumina membranes were used for this method [13].

Mainly there are two types of semiconductor materials-Organic and Inorganic. Organic semiconductor materials possess more ability to absorb and release light energy and this property was already utilised in laser printers, photoconductors etc. Thin film based organic semiconductor are more preferable for practical applications and it is more compatible for photodetectors and photovoltaic cells. Inorganic materials also emerge as an efficient alterative in this regard recently. It also shows high compatibility in all types of luminescent devices including QDLED. The detailed study of Organic and Inorganic semiconductor devices are discussed below.

2.2 An Overview to the p–n Junction Diode

The detailed study of the principle of p–n junction diode is described here. In any p–n junction device, the material with high electron carriers is said to be n-type semiconductor material and if the hole carriers are dominant than the electron, then it is said to be p-type semiconductor material [14]. All materials consist of a valance band and a conduction band. Valance band consists of free electrons as the charge carriers, which is excited to the conduction band due to external effects such as temperature, light, voltage applied between the junction diode etc. by leaving a hole in the valance band. This excited electron emits spontaneously by releasing the extra energy in the form of light and recombine with the hole to get an equilibrium and this is called radiative recombination [15]. The recombination can also be non-radiative in which the energy difference won't reflect as light emission. The emitted energy is equivalent to the energy acquired by an atom which is excited from valance band to the conduction band. A fermi energy level exists in the middle of the valance band and conduction band which moves with respect to the number of charge carriers present in each band. For intrinsic semiconductor materials, the fermi level lies at exact middle point of the band by Eq. (2.1), since the effective mass of electrons and holes are the same here.

$$E_F = \frac{E_C + E_V}{2} + \frac{3}{4}kTln\frac{m_e}{m_h} \tag{2.1}$$

where E_C—band gap between conduction band and the fermi level, E_V—band gap between valance band and the fermi level, m_e and m_h are the effective masses of electron and hole respectively. A p-type and n-type material can be made by doping a material with less valance electrons (acceptors) and excess valance electrons (donors) than the intrinsic material. If the material is doped with donors, the effective mass of electrons in the valance band will increase, so that the fermi level will be shifted towards the conduction band. If the material is doped with acceptors, the number of hole carriers will increase in the conduction band and the fermi level moves towards to valance band. The conductivity of n-type material is more than the p-type material, since the mobility of electron is greater than the hole.

2.2.1 Organic Material Devices

As it is mentioned above, the organic materials are the ideal materials for applications like photodiodes, solar cells, thin film transistors, LED etc.

The OLED was first developed in France in 1950s and the same was demonstrated in 1987 by the group of Tang and Van Slyke at Kodak from thin organic layers and they reported an external efficiency (ratio of number of photons emitted to the number of electrons injected) of about > 2%. The blue and green OLED was first explored more

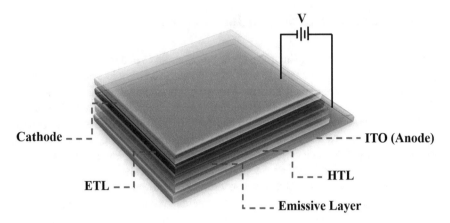

Fig. 2.1 Basic structure of OLED

than the red. Because the red OLED showed less efficiency than the blue and greens [16]. The OLED operates on the same principle of a regular p–n junction diode. The emissive layer (EL) of organic compound (situated in the middle of device) emit light as a response to an electric current [17] (Fig. 2.1).

On either part of the emissive layer, there is a cathode and anode layer. Below the cathode layer, there is an electron transport layer (ETL), which transports electrons from the cathode to the emissive layer. Same time, there is a hole transport layer (HTL) below the emissive layer, which helps to inject the holes from the anode to the emissive layer. When a voltage is applied through the device, the electrons will move from the fermi level of the cathode to the lowest unoccupied molecular orbitals (LUMO) of the emissive layer through ETL and the hole will move from the fermi level of anode to the highest occupied molecular orbitals (HOMO) of the emissive layer through HTL and the recombination takes place in the emissive layer, which produces light [18]. The color of the emitted light mainly depends on the emissive layer or organic layer, which can be modified by changing the chemical composition. The charge injection layer of OLED determines the efficiency of the device [19]. The lifetime of the OLED also depends on the efficiency of the injection process [20]. While discussing about the efficiency of a device it may include internal quantum efficiency, light extraction efficiency and electrical characteristics etc. The internal quantum efficiency is nothing but the ratio of number of photons generated within the structure to the number of electrons injected [21]. This can be improved by fabricating multilayer structures [22], doping with fluorescent dyes [23], improving carrier-injection layers [24] and using phosphorescent harvesters [25]. Figure 2.2 shows the emission spectra of three layered EL devices fabricated by Adachi et al. using anthracene, coronene and perylene. It shows three emission maxima in the spectra at 420 nm (purple blue), 500 nm (green) and 600 nm (orange) corresponding to anthracene, coronene and perylene. These emissions are produced by the singlet excited states by the recombination of electrons and holes in the emitting layer [22].

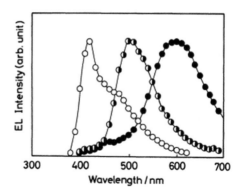

In phosphorescent dyes, emissions may result from both the singlet and the triplet states, which improve the light emission efficiency [25]. The EL colors can also be tuned by doping the multi-layered structure with different dopants under different concentrations [23].

The Organic light emitting devices utilizes small organic molecule or polymers as the light emitting semiconductor material. OLEDs are widely used in many digital displays such as televisions, computer monitors, portable systems etc. However, the low light extraction efficiency of OLEDs leads the researchers to develop buckling structure, which can improve the efficiency OLEDs more than the normal one [21, 26]. Koo et al. demonstrated a buckling structure with broad distribution and directional randomness, which can enhance the light extraction efficiency without introducing spectral changes and directionality [27]. It is found that, the OLEDs with buckled structure showed improved current and power efficiency and an electroluminescence spectrum enhanced by a factor of two across the entire visible wavelength regime. The buckled structure can be formed spontaneously on the thin metallic film and it can be used for getting full color and white color OLEDs. Generally, Indium Tin Oxide (ITO) is used as anode layer or organic layer in OLEDs. But the refractive index of ITO is about 1.7–2 and the 40–60% of emitted lights are trapped in this layer as wave guide modes. To overcome this, the Braggs diffraction grating and low index grating can be done to extract the ITO or organic modes [28]. The 2-D layer structure like Photonic crystal (PC) can increase the efficiency of light extraction in OLEDs. Fujita et al. fabricated the PC structures in organic and electrode layers to extract the light in waveguide mode and they observed that, the OLEDs efficiency in terms of spectrally integrated intensity was improved to 20% and the peak intensity of forward emitting light was improved up to 130% [29].

Organic photodiode (OPD) was first demonstrated by Kudo and Moriizumi in 1981 [30, 31]. After that there reports on the fabrication of OPD s which are sensitive to different colors like Blue [32, 33], Green [34] and Red [35] OPDs are found by improving the efficiency of the device. Solar cells based on organic semiconductor materials also strongly made their foundation in device technology. Many researchers are trying to improve the power conversion efficiency and stability of the device [36–41].

2.2.2 *Inorganic Devices*

Although Organic based LEDs, LASERS and photodiodes possess superior proper-ties, their low stability is the biggest challenge and hence needs efficient alternatives. Hence, currently researchers are trying to develop novel inorganic materials with superior optical properties for functional devices because of their high stability, shelf life and tuneable functionality and properties. Current trend in this regard are quantum dot based optical devices rather than quantum well and quantum wire.

2.2.2.1 Quantum Dot Devices

Quantum confinement effect is the most unique property of a QDs, which modifies the density of states near the band edges. The density of quantum states as a function of energy for a QDs lies between discrete atomic and continuous bulk materials (Fig. 2.3). When the band gap energy of the crystal is more and exceeds $k_B T$ (k_B-Boltzmann constant, T-temperature), and the corresponding particle size is small leading to quantum confinement effect. If the energy differences become $> k_B T$, it restricts the mobility of electrons and holes in the crystal. In QDs, discrete energy states can be observed due to the presence of small atoms and each energy level can exhibit wave functions. The solutions of each wave functions are very similar

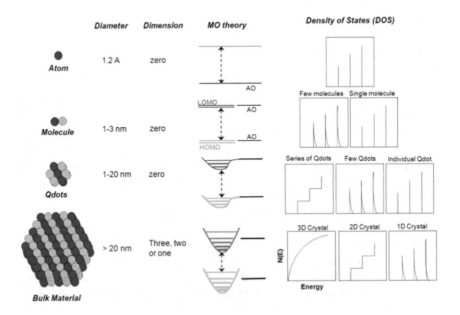

Fig. 2.3 Schematic shows the density of quantum states as a function of energy with respect to the discrete atomic and continuous bulk materials

to the electron-nucleus bond present in an atom. Therefore, a sharp emission can be possible in QDs.

Optical applications of QDs can be enhanced by doping with any other transition metal [42–49]. It perturbs the band structure of the material within the band gap by creating their own quantum states. This dopant can be self-autoionized due to the confinement effect. The doping of a material to any QDs, can increase the confinement energy due to the smaller size. If the energy is exceeding the columbic interaction energy between electron and holes, autoionization can occur. The QDs also possess' high surface to volume ratio, which will enhance the optical properties of them such as high luminescence, high quantum efficiency etc. The optical properties of a material also depend on the excitons (electron hole pair). The energy of an exciton is less than the band gap energy and it behaves like a hydrogen atom. The distance between electron and hole is said to be exciton Bohr radius (a_B). If m_h and m_e are the masses of holes and electrons respectively, then the exciton Bohr radius can be written as

$$a_B = \frac{\hbar^2 \in}{e^2} \left(\frac{1}{m_h} + \frac{1}{m_e} \right) \qquad (2.2)$$

The relation between the size of the QD with exciton Bohr radius will tell, whether the QD is under strong confinement region ($D < 2a_B$), intermediate confinement region ($2a_B < D < 4a_B$ or weak confinement region ($D > 4a_B$), where D is the crystallite size of the particle. The stability of the QD devices can increase by over coating inorganic material with a wider band gap (as shell) such as ZnS, CdS etc.

Nowadays, scientists are mainly focusing on QDs as they offer high device application potential due to its confinement effect. QDs have many remarkable properties and it can emit visible light from blue to red colors by tuning their size to several nanometers and it shows high color purity, solution processability and stability [50]. As compared to Organic LED (OLED), the QD LED has more color gamut (~> 90%) because of its narrow wavelength emission in both photoluminescence (PL) and electroluminescence (EL) [51] and it may be possible to obtain all three primary colors with the same composition by changing the particle sizes. By incorporating three primary color QDs between the ETL and HTL, a white LED was fabricated [52], which has an EQE of 0.36% at 10 V applied bias. It is observed that, the QD LED can emit light under both forward and reverse bias [53]. This is possible due to the different rates of electron and hole injection, different carrier mobility of electron and holes in the corresponding transporting layer and uniformity of the film layer. Previously, some researchers used cadmium selenide QDs for electroluminescence devices, which can emit light in the wavelength range 520–610 nm. But the EQE of such device was found to be very less (about 0.01%) [54, 55]. The fabrication of a bilayer LEDs with organically capped CdSe (CdS) core/shell type semiconductor nanocrystals and electroluminescent semiconducting polymer [poly (p-phenylnevinylene)] shows EQE upto 0.22% by emitting red to green lights [56]. The EQE of device can also be increased by incorporating organic poly phenylene vinylene with CdSe nanocrystals and by making thin films of CdSe-ZnS core–shell nanocrystals [57, 58]. Later in 2003, Steckel et al. demonstrated a device using PbSe colloidal QDs. In their work, they tuned the wavelength of electroluminescence from 1330 to 1560 nm by changing the QD size [59]. Another group reported the fabrication of a trilayer hybrid polymer QDLED by sandwiching a CdSe-ZnS core–shell QD layer between films of polyvinylecarbazol (PVK) and an oxadiazole derivative (butyle PBD) using spin coating technique. It has shown the EQE close to 0.2% [60]. M. C. Jean et al. demonstrated a hybrid inorganic/organic light emitting device composed of a CdSe-ZnS core–shell semiconductor QD emissive layer sandwiched between p-type NiO and tris-(8-hydroxyquinoline) aluminium (Alq3) as HTL and ETL and they achieved 0.18% of EQE by tuning the resistivity of the NiO layer to balance the electron and hole densities at QD sites [61]. Based on $Cd_{1-x}Zn_xS@ZnS$, a blue LED was demonstrated by K. B Wan et al. and they found the EQE of (0.1– 0.3%) [62]. An electroluminescence from a mixed monolayer of red, green and blue emitting QD LED in hybrid organic/inorganic structure was demonstrated by Anikeeva et al. in their laboratory. In their work, they synthesized three types of colloidal QDs: CdSe/ZnS core shell QDs (for red color), ZnSe/CdSe alloyed cores overcoated with a shell of ZnS (for green color) and Zn-CdS alloyed QDs (for blue color). These three QDs solution shown PL peaks at 620 nm, 540 nm and 440 nm respectively. By mixing all three QDs solution, they prepared white QD LED [52]. The normalized EL spectra and EQE of QDLEDs are shown in Fig. 2.4.

From all the reviews we discussed above, it is found that, EL devices can be fabricated often with cadmium-based materials, because of the high luminescent properties of CdS, CdSe etc. Cadmium based materials also show good external quantum efficiency (EQE) and lifetime in devices, especially in red and green emitting devices (reported as > 20%) [63, 64]. But the toxicity of cadmium-based materials and their possible long term environmental impact leave them at the back seat for large scale production. To overcome these, and to decrease the environmental impact, researchers are focusing on cadmium free QDLED and it is reported that, the EQE for red emission is high (~21.4%) [65] compared with green (~15.6%) and blue (~12.4%) emissions [66]. R. Tatsuya reported the outstanding EL properties of QD LED using cadmium free QDs. They got EQE as 14.7% and 10.7% for cadmium free QD LED (ZnSe QD LED) emitting at the wavelength 428 nm and 445 nm respectively [51]. C. Weiran et al. synthesised high photoluminescence quantum yield QDs with ZnSe outer shell. This ZnSe based QD LED show a lifetime T_{50} over 2 million hours [67].

To obtain a broad band emission spectrum using a single material is not an easy job. Therefore, the full color and effective QD display technology is still a challenging one in research area. As per the theory, the emission at higher wavelength is possible by increasing the size of the QD. However, control the stability of nanoparticle (NP), crystal quality etc. are rather difficult.

Solar cell device is another application of colloidal QD material. It contains photons with energy ranging from 0.5 to 3.5 eV. There are mainly four processes occur in solar cells (i) absorption of light and exciton formation (ii) exciton diffusion (iii) charge separation and (iv) charge transportation.

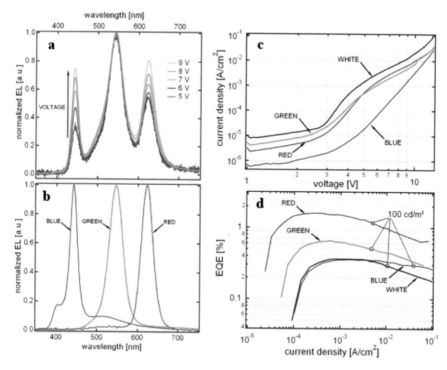

Fig. 2.4 **a** Normalized EL spectra of a white QD-LED for a set of increasing applied voltages. The relative intensities of red and blue QD spectral components increase in comparison to the green QD component at higher biases. **b** Normalized EL spectra of red, green, and blue monochrome QD-LEDs (red, green, and blue lines, respectively). Current–voltage characteristics **c** and external electroluminescence quantum efficiency **d** measured for red, green, blue, and white QD-LEDs labeled with red, green, blue and black lines, respectively. The circled data points indicate device brightness of 100 cd/m^2. Reprinted with permission from Ref. [52]. Copyright (2007) American Chemical Society

It was found that the combination of inorganic and organic material can improve the efficiency of solar cells. Such types of solar cells using CdSe [68], PbS [69], PbSe [70], ZnO [71], TiO$_2$ [72], CuInS$_2$ [73], CuInSe$_2$ [74] are reported. The QD can absorb all entire gamut of sunlight from UV to Visible to IR due to its confinement property and it can generate multiple excitons with a single photon. QD solar cell has high potential to increase the efficiency of solar photon conversion up to about 66% by utilizing high photogenerated carriers. QD solar cells are broadly classified into three; QD sensitized, QD dispersed and P-I-N solar cell with QDs array [75]. QD sensitized solar cell works based on the dye sensitization of nanocrystalline TiO$_2$ layers [76, 77]. Upon photoexcitation of the dye molecules, the electrons are very efficiently injected from the excited state of the dye into the conduction band of the TiO$_2$, affecting charge separation and producing photovoltaic effect. In QD dispersed solar cells, QDs will form a junction with organic semiconductor polymers and upon photoexcitation, the photogenerated holes are injected into the polymer phase. In P-I-N solar cells, the QDs are formed into an orders 3-D array with inter QD spacing such that strong electronic coupling occurs and minibands are formed to allow long-range electron transport.

2.2.2.2 Quantum Well Devices

A we discussed earlier, the quantum well is a 2D nanostructure, in which particle can free to move in 2-dimension and confined in 1-dimention. A quantum well structure can be made by sandwiching a thin layer of semiconductor material between two layer of other semiconductor material. Molecular beam epitaxy (MBE), Metalorganic chemical vapour deposition (MOCVD), chemical beam epitaxy (CBE) etc. can be used for fabricating a quantum well material [78]. Semiconductor quantum wells are used to fabricate lasers [79, 80], photodetectors [81, 82], LED [83, 84], solar cells [85] and hybrid devices [86] etc. Quantum wells are also used as absorbers in semiconductor saturable absorber mirrors (SESAMs) [87] and in electro absorption modulators [88, 89]. Former devices are used for the generation of ultrashort pulses by passive mode locking of lasers and latter is used for controlling the intensity of the laser beam.

Quantum well lasers are an attractive technology in semiconductor devices and the first quantum well lasers operated at a wavelength about 800 nm and now it can be fabricated from the visible to infrared region. Its properties can be changed by varying the width of the quantum well. Therefore, quantum well lasers can be considered somewhat better than the conventional lasers. While moving to the quantum well solar cells, a promising approach for the next generation photovoltaic technology have received great attention in research field. I. sayed and S M Bedair discussed about the advantages and challenges of growing quantum wells in the unintentionally doped region of p-i-n solar cells. They focused on, 1.1–1.3 eV strain-balanced InGaAs/GaAsP, 1.6–1.8 eV strain-balanced and lattice-matched InGaAsP/InGaP, and >2.1 eV strained InGaN/GaN quantum well solar cells, including optimization of the quantum well growth conditions and improving of the structure [90]. For each material, the device performance, thickness constraints, bandgap tunability, and carrier transport limitations are discussed. R. E. Welser et al. also designed and demonstrated high efficiency (>26%) single junction quantum well solar cells by employing thin strained super lattices and a heterojunction emitter [85]. A multiple quantum well solar cells was investigated by T. Kasidit et al. using the effective mobility model [91]. Both experimental and simulation study has shown the role of electron and hole densities in carrier collection process. Hybrid solar cells also shows the high efficiency in solar thermal powerplants [92].

2.2.2.3 Quantum Wire Devices

In quantum wires, electrons are confined in 2-dimension and are free to move only in one-dimension. It appears like wires or tubes with diameters in the nanometre range. In quantum wires, electron transport is possible only when the energy is less than the fermi energy. Very few electrons are scattered and it attain higher mobilities while travelling through the quantum wire. Therefore, lower dimension devices could be powered by far less current and attain much attention in optical application. Like QDs and quantum wells, quantum wires also offer promising applications in lasers, diodes, transistors, photovoltaics etc. Out of these, Quantum wire laser devices are the most important applications where the special optical properties of 1-D structures can be exploited. E. Kapon et al. fabricated a single quantum wire GaAs/AlGaAs injection laser using organometallic chemical vapour deposition on V-grooved GaAs substrates. Results shown the average and lowest threshold current as 4.4 mA and 3.5 mA respectively at room temperature [93]. The performance of quantum wire laser was improved by S. Tiwary et al. by employing multiple quantum wire laser configuration [94]. They constructed a structure of $Ga_{1-x}In_xAs/Ga_{1-x}Al_xAs$ semiconductor material with a minimum threshold current 188 μA and maximum powers of about 50 μW in continuous multimode operation at wavelength of 980 nm. The internal quantum efficiency (IQE) obtained as 83%. Demonstration of single quantum wire laser was done by H. Yuhei et al. and they got the threshold power of 5 mW at 5 K temperature [95]. A one step chemical vapour deposition process was described by J. C. Heon et al. for making quantum wire lasers based on Al-Ga-N system. They obtained a novel quantum wire in optical fibre nanostructures. This leads to the first GaN based quantum wire UV lasers with low threshold [96]. A narrow silicon quantum wire transistor can be fabricated using different process such as electron beam lithography, anisotropic dry etching thermal oxidation etc. with embedded silicon dioxide. It can be observed that a quantized conductance staps originates below 4.2 K temperature [97].

Quantum wire photodetector is also one of the emerging applications in semiconductor nanostructures. Some literature described an infra-red photodetector using quantum wire. D. Biswajit and S. Pavan discussed a quantum wire infra-red detector in long wavelength region based on inter-sub-band transitions in semiconductor quantum wire. They also discussed operational principle, design procedure and implementation of the quantum wire detector [98]. An InGaAs quantum wire photodetector was fabricated on (001) axis InP substrate by molecular beam epitaxy method which showed unique polarized photoresponse [99]. The performance of quantum wire infra-red photodetector was studied by A. Nasr in which it was noticed a gradual increasing of photocurrent from the numerical results [100].

2.3   Conclusion

In Summary, the new era in semiconductor industry will be dominated by the quantum structured devices. Especially the optical devices, starting from light emitting ones (LED and LASER) photovoltaics and optical sensors/ detectors, of the next generation will have atleast one component in which the materials will be in confinement regime and more and more novel materials and their effective combinations are being developed around the globe to improve the efficiency shelf life and stability. Efforts are being made to replace toxic lead, Cd and As based structures with more environmentally friendly materials like ZnS, ZnSe etc. and large scope exists in this field in the nearby future.

Semiconductor QDs are also an excellent material for the cellular and bio imaging application due to the several properties of the material, which is discussed in detail in the next chapter.

References

1. H.S. Mansur, Quantum dots and nanocomposites. Wiley Interdiscip. Rev. Nanomedicine Nanobiotechnology **2**, 113–129 (2010)
2. L. Jacak, P. Hawrylak, A. Wojs, *Quantum Dots* (Springer Science & Business Media, 2013).
3. J.G. Lu et al., Self-assembled ZnO quantum dots with tunable optical properties. Appl. Phys. Lett. **89**, 23122 (2006)
4. Y. Yang et al., High-efficiency light-emitting devices based on quantum dots with tailored nanostructures. Nat. Photonics **9**, 259–266 (2015)
5. F. Yuan et al., Engineering triangular carbon quantum dots with unprecedented narrow bandwidth emission for multicolored LEDs. Nat. Commun. **9**, 1–11 (2018)
6. F. Yuan et al., Highly efficient and stable white LEDs based on pure red narrow bandwidth emission triangular carbon quantum dots for wide-color gamut backlight displays. Nano Res. **12**, 1669–1674 (2019)
7. Z. Wang, X. Dong, S. Zhou, Z. Xie, Z. Zalevsky, Ultra-narrow-bandwidth graphene quantum dots for superresolved spectral and spatial sensing. NPG Asia Mater. **13**, 1–13 (2021)
8. G. Bai, M. Tsang, J. Hao, Tuning the luminescence of phosphors: beyond conventional chemical method. Adv. Opt. Mater. **3**, 431–462 (2015)
9. Y. Ma, Y. Zhang, W.Y. William, Near infrared emitting quantum dots: synthesis, luminescence properties and applications. J. Mater. Chem. C **7**, 13662–13679 (2019)
10. B. Ghosh, N. Shirahata, Colloidal silicon quantum dots: synthesis and luminescence tuning from the near-UV to the near-IR range. Sci. Technol. Adv. Mater. **15**, 14207 (2014)
11. L. Esaki, R. Tsu, Superlattice and negative differential conductivity in semiconductors. IBM J. Res. Dev. **14**, 61–65 (1970)
12. M. Kuno, Introductory nanoscience: Physical and chemical concepts. MRS Bull. **37**, 169–170 (2012)
13. W. Zaghdoudi, T. Dammak, H. ElHouichet, R. Chtourou, Synthesis, optical and structural properties of quantum-wells crystals grown into porous alumina. Superlattices Microstruct. **71**, 117–123 (2014)
14. A. E. Dixon, Review of solid state physics, in *Solar Energy Conversion* (Elsevier, 1979), pp. 773–784
15. W.P. Dumke, Spontaneous radiative recombination in semiconductors. Phys. Rev. **105**, 139 (1957)

16. C.W. Tang, S.A. VanSlyke, Organic electroluminescent diodes. Appl. Phys. Lett. **51**, 913–915 (1987)
17. A. Khazanchi, A. Kanwar, L. Saluja, A. Damara, V. Damara, OLED: a new display technology. Int. J. Eng. Comput. Sci. **1**, 75–84 (2012)
18. J.C. Scott, G.G. Malliaras, Charge injection and recombination at the metal–organic interface. Chem. Phys. Lett. **299**, 115–119 (1999)
19. G.G. Malliaras, J.R. Salem, P.J. Brock, C. Scott, Electrical characteristics and efficiency of single-layer organic light-emitting diodes. Phys. Rev. B **58**, R13411 (1998)
20. S.A. Carter, M. Angelopoulos, S. Karg, P.J. Brock, J.C. Scott, Polymeric anodes for improved polymer light-emitting diode performance. Appl. Phys. Lett. **70**, 2067–2069 (1997)
21. S.R. Forrest, D.D.C. Bradley, M.E. Thompson, Measuring the Efficiency of Organic Light-Emitting Devices. Adv. Mater. **15**, 1043–1048 (2003)
22. C. Adachi, S. Tokito, T. Tsutsui, S. Saito, Electroluminescence in organic films with three-layer structure. Jpn. J. Appl. Phys. **27**, L269 (1988)
23. C.W. Tang, S.A. VanSlyke, C.H. Chen, Electroluminescence of doped organic thin films. J. Appl. Phys. **65**, 3610–3616 (1989)
24. A. Elschner et al., PEDT/PSS for efficient hole-injection in hybrid organic light-emitting diodes. Synth. Met. **111**, 139–143 (2000)
25. M.A. Baldo et al., Highly efficient phosphorescent emission from organic electroluminescent devices. Nature **395**, 151–154 (1998)
26. A. Chutinan, K. Ishihara, T. Asano, M. Fujita, S. Noda, Theoretical analysis on light-extraction efficiency of organic light-emitting diodes using FDTD and mode-expansion methods. Org. Electron. **6**, 3–9 (2005)
27. W.H. Koo et al., Light extraction from organic light-emitting diodes enhanced by spontaneously formed buckles. Nat. Photonics **4**, 222 (2010)
28. J.M. Ziebarth, A.K. Saafir, S. Fan, M.D. McGehee, Extracting light from polymer light-emitting diodes using stamped bragg gratings. Adv. Funct. Mater. **14**, 451–456 (2004)
29. M. Fujita et al., Optical and electrical characteristics of organic light-emitting diodes with two-dimensional photonic crystals in organic/electrode layers. Jpn. J. Appl. Phys. **44**, 3669 (2005)
30. K. Kudo, T. Moriizumi, Photoelectric pn junction cells using organic dyes. Jpn. J. Appl. Phys. **20**, L553 (1981)
31. K. Kudo, T. Moriizumi, Spectrum-controllable color sensors using organic dyes. Appl. Phys. Lett. **39**, 609–611 (1981)
32. T. Fukuda et al., Improved optical-to-electrical conversion efficiency by doping silole derivative with low ionization potential. Phys. status solid **209**, 2324–2329 (2012)
33. T. Fukuda, S. Kimura, R. Kobayashi, A. Furube, Ultrafast study of charge generation and device performance of a silole-doped fluorene-mixed layer for blue-sensitive organic photoconductive devices. Phys. status solid **210**, 2674–2682 (2013)
34. T. Fukuda, S. Kimura, Z. Honda, N. Kamata, Solution-processed green-sensitive organic photoconductive device using rhodamine 6G. Mol. Cryst. Liq. Cryst. **566**, 67–74 (2012)
35. M. Kaneko, T. Taneda, T. Tsukagawa, H. Kajii, Y. Ohmori, Fast response of organic photodetectors utilizing multilayered metal-phthalocyanine thin films. Jpn. J. Appl. Phys. **42**, 2523 (2003)
36. A. Wadsworth et al., Critical review of the molecular design progress in non-fullerene electron acceptors towards commercially viable organic solar cells. Chem. Soc. Rev. **48**, 1596–1625 (2019)
37. S. Zhang et al., Interface engineering via phthalocyanine decoration of perovskite solar cells with high efficiency and stability. J. Power Sources **438**, 226987 (2019)
38. K. Rakstys, C. Igci, M.K. Nazeeruddin, Efficiency vs. stability: dopant-free hole transporting materials towards stabilized perovskite solar cells. Chem. Sci. **10**, 6748–6769 (2019)
39. C. Yao et al., Trifluoromethyl group-modified non-fullerene acceptor toward improved power conversion efficiency over 13% in polymer solar cells. ACS Appl. Mater. Interfaces **12**, 11543–11550 (2020)

40. F. Wang et al., Interface dipole induced field-effect passivation for achieving 21.7% efficiency and stable perovskite Solar cells. Adv. Funct. Mater. 2008052 (2020)
41. B. Subramanyam, P.C. Mahakul, K. Sa, J. Raiguru, P. Mahanandia, Investigation of improvement in stability and power conversion efficiency of organic solar cells fabricated by incorporating carbon nanostructures in device architecture. J. Phys. Mater. 3, 45004 (2020)
42. J.-U. Kim, Y.K. Kim, H. Yang, Reverse micelle-derived Cu-doped Zn1−xCdxS quantum dots and their core/shell structure. J. Colloid Interface Sci. 341, 59–63 (2010)
43. Y. Yang, O. Chen, A. Angerhofer, Y.C. Cao, Radial-position-controlled doping in CdS/ZnS core/shell nanocrystals. J. Am. Chem. Soc. 128, 12428–12429 (2006)
44. H. Yang, S. Santra, P.H. Holloway, Syntheses and applications of Mn-doped II-VI semiconductor nanocrystals. J. Nanosci. Nanotechnol. 5, 1364–1375 (2005)
45. A. Murugadoss, A. Chattopadhyay, Tuning photoluminescence of ZnS nanoparticles by silver. Bull. Mater. Sci. 31, 533–539 (2008)
46. S. Hou et al., Photoluminescence and XPS investigations of Cu^{2+}-doped ZnS quantum dots capped with polyvinylpyrrolidone. Phys. status solidi 246, 2333–2336 (2009)
47. R. Beaulac, L. Schneider, P.I. Archer, G. Bacher, D.R. Gamelin, Light-induced spontaneous magnetization in doped colloidal quantum dots. Science 325, 973–976 (2009)
48. Z. Yu-Hong et al., Cr-doped InAs self-organized diluted magnetic quantum dots with room-temperature ferromagnetism. Chinese Phys. Lett. 24, 2118 (2007)
49. R.N. Bhargava, Doped nanocrystalline materials—physics and applications. J. Lumin. 70, 85–94 (1996)
50. V. Wood, V. Bulović, Colloidal quantum dot light-emitting devices. Nano Rev. 1, 5202 (2010)
51. T. Ryowa et al., High-efficiency quantum dot light-emitting diodes with blue cadmium-free quantum dots. J. Soc. Inf. Disp. 28, 401–409 (2020)
52. P.O. Anikeeva, J.E. Halpert, M.G. Bawendi, V. Bulović, Electroluminescence from a mixed red–green–blue colloidal quantum dot monolayer. Nano Lett. 7, 2196–2200 (2007)
53. E. O'Connor et al., Near-infrared electroluminescent devices based on colloidal HgTe quantum dot arrays. Appl. Phys. Lett. 86, 201114 (2005)
54. V.L. Colvin, M.C. Schlamp, A.P. Alivisatos, Light-emitting diodes made from cadmium selenide nanocrystals and a semiconducting polymer. Nature 370, 354–357 (1994)
55. B.O. Dabbousi, M.G. Bawendi, O. Onitsuka, M.F. Rubner, Electroluminescence from CdSe quantum-dot/polymer composites. Appl. Phys. Lett. 66, 1316–1318 (1995)
56. M.C. Schlamp, X. Peng, A. Alivisatos, Improved efficiencies in light emitting diodes made with CdSe (CdS) core/shell type nanocrystals and a semiconducting polymer. J. Appl. Phys. 82, 5837–5842 (1997)
57. H. Mattoussi et al., Composite thin films of CdSe nanocrystals and a surface passivating/electron transporting block copolymer: Correlations between film microstructure by transmission electron microscopy and electroluminescence. J. Appl. Phys. 86, 4390–4399 (1999)
58. H. Mattoussi et al., Self-assembly of CdSe− ZnS quantum dot bioconjugates using an engineered recombinant protein. J. Am. Chem. Soc. 122, 12142–12150 (2000)
59. J.S. Steckel, S. Coe-Sullivan, V. Bulović, M.G. Bawendi, 1.3 μm to 1.55 μm tunable electroluminescence from PbSe quantum dots embedded within an organic device. Adv. Mater. 15, 1862–1866 (2003)
60. S. Chaudhary, M. Ozkan, W.C.W. Chan, Trilayer hybrid polymer-quantum dot light-emitting diodes. Appl. Phys. Lett. 84, 2925–2927 (2004)
61. J.-M. Caruge, J.E. Halpert, V. Bulović, M.G. Bawendi, NiO as an inorganic hole-transporting layer in quantum-dot light-emitting devices. Nano Lett. 6, 2991–2994 (2006)
62. W.K. Bae et al., Deep blue light-emitting diodes based on Cd1−xZnxS@ ZnS quantum dots. Nanotechnology 20, 75202 (2009)
63. X. Dai et al., Solution-processed, high-performance light-emitting diodes based on quantum dots. Nature 515, 96–99 (2014)
64. K.P. Acharya et al., High efficiency quantum dot light emitting diodes from positive aging. Nanoscale 9, 14451–14457 (2017)

65. Y.-H. Won et al., Highly efficient and stable InP/ZnSe/ZnS quantum dot light-emitting diodes. Nature **575**, 634–638 (2019)
66. C. Ippen et al., High efficiency heavy metal free QD-LEDs for next generation displays. J. Soc. Inf. Disp. **27**, 338–346 (2019)
67. W. Cao et al., Highly stable QLEDs with improved hole injection via quantum dot structure tailoring. Nat. Commun. **9**, 1–6 (2018)
68. L. Han et al., Synthesis of high quality zinc-blende CdSe nanocrystals and their application in hybrid solar cells. Nanotechnology **17**, 4736 (2006)
69. P. Michler et al., Quantum correlation among photons from a single quantum dot at room temperature. Nature **406**, 968–970 (2000)
70. D. Cui et al., Harvest of near infrared light in PbSe nanocrystal-polymer hybrid photovoltaic cells. Appl. Phys. Lett. **88**, 183111 (2006)
71. D.C. Olson, J. Piris, R.T. Collins, S.E. Shaheen, D.S. Ginley, Hybrid photovoltaic devices of polymer and ZnO nanofiber composites. Thin Solid Films **496**, 26–29 (2006)
72. C.C. Oey et al., Polymer–TiO_2 solar cells: TiO_2 interconnected network for improved cell performance. Nanotechnology **17**, 706 (2006)
73. B. O'Regan, D.T. Schwartz, S.M. Zakeeruddin, M. Grätzel, Electrodeposited nanocomposite n–p heterojunctions for solid-state dye-sensitized photovoltaics. Adv. Mater. **12**, 1263–1267 (2000)
74. S. Bereznev, I. Konovalov, A. Öpik, J. Kois, Hybrid $CuInS_2$/polypyrrole and $CuInS_2$/poly (3, 4-ethylenedioxythiophene) photovoltaic structures. Synth. Met. **152**, 81–84 (2005)
75. A.J. Nozik, Quantum dot solar cells. Phys. E Low-Dimensional Syst. Nanostructures **14**, 115–120 (2002)
76. J.E. Moser, P. Bonnóte, M. Grätzel, Molecular photovoltaics. Coord. Chem. Rev. **171**, 245–250 (1998)
77. A. Hagfeldt, M. Grätzel, Molecular photovoltaics. Acc. Chem. Res. **33**, 269–277 (2000)
78. P.K. Bhattacharya, N.K. Dutta, Quantum well optical devices and materials. Annu. Rev. Mater. Sci. **23**, 79–123 (1993)
79. P.S. Zory, Optical gain in III–V bulk and quantum well semiconductors, in *Quantum Well Lasers* (Academic, 1993),pp. 17–96
80. B. Zhao, A. Yariv, Quantum well semiconductor lasers, in *Semiconductor Lasers I* (Elsevier, 1999), pp. 1–121
81. J.C. Cao, H.C. Liu, Terahertz semiconductor quantum well photodetectors, in *Semiconductors and Semimetals*, vol. 84, (Elsevier, 2011), pp. 195–242
82. A. Caria et al., Excitation Intensity and Temperature-Dependent Performance of InGaN/GaN Multiple Quantum Wells Photodetectors. Electronics **9**, 1840 (2020)
83. A.K. Viswanath, Surface and interfacial recombination in semiconductors. Handb. surfaces interfaces Mater. **1**, 217–284 (2001)
84. L. Cheng et al., Multiple-quantum-well perovskites for high-performance light-emitting diodes. Adv. Mater. **32**, 1904163 (2020)
85. R.E. Welser et al., Design and demonstration of high-efficiency quantum well solar cells employing thin strained superlattices. Sci. Rep. **9**, 1–10 (2019)
86. F. Vigneau et al., Germanium quantum-Well Josephson field-effect transistors and interferometers. Nano Lett. **19**, 1023–1027 (2019)
87. U. Keller et al., Semiconductor saturable absorber mirrors (SESAM's) for femtosecond to nanosecond pulse generation in solid-state lasers. IEEE J. Sel. Top. Quantum Electron. **2**, 435–453 (1996)
88. E. Lach, K. Schuh, M. Schmidt, Application of electroabsorption modulators for high-speed transmission systems, in *Ultrahigh-Speed Optical Transmission Technology* (Springer, 2005), pp. 347–377
89. B. Pezeshki, S.M. Lord, T.B. Boykin, J.S. Harris Jr., GaAs/AlAs quantum wells for electroabsorption modulators. Appl. Phys. Lett. **60**, 2779–2781 (1992)
90. I. Sayed, S.M. Bedair, Quantum well solar cells: principles, recent progress, and potential. IEEE J. Photovoltaics **9**, 402–423 (2019)

91. K. Toprasertpong, S.M. Goodnick, Y. Nakano, M. Sugiyama, Effective mobility for sequential carrier transport in multiple quantum well structures. Phys. Rev. B **96**, 75441 (2017)
92. G. Moses et al., InGaN/GaN multi-quantum-well solar cells under high solar concentration and elevated temperatures for hybrid solar thermal-photovoltaic power plants. Prog. Photovoltaics Res. Appl. **28**, 1167–1174 (2020)
93. E. Kapon, S. Simhony, R. Bhat, D.M. Hwang, Single quantum wire semiconductor lasers. Appl. Phys. Lett. **55**, 2715–2717 (1989)
94. S. Tiwari et al., High efficiency and low threshold current strained V-groove quantum-wire lasers. Appl. Phys. Lett. **64**, 3536–3538 (1994)
95. Y. Hayamizu et al., Lasing from a single-quantum wire. Appl. Phys. Lett. **81**, 4937–4939 (2002)
96. H.-J. Choi et al., Self-organized GaN quantum wire UV lasers. J. Phys. Chem. B **107**, 8721–8725 (2003)
97. M. Je, S. Han, I. Kim, H. Shin, A silicon quantum wire transistor with one-dimensional subband effects. Solid. State. Electron. **44**, 2207–2212 (2000)
98. B. Das, P. Singaraju, Novel quantum wire infrared photodetectors. Infrared Phys. Technol. **46**, 209–218 (2005)
99. C.L. Tsai, K.Y. Cheng, S.T. Chou, S.Y. Lin, InGaAs quantum wire infrared photodetector. Appl. Phys. Lett. **91**, 181105 (2007)
100. A. Nasr, Performance of quantum wire infrared photodetectors under illumination conditions. Opt. Laser Technol. **41**, 871–876 (2009)

Chapter 3
Semiconductor Quantum Dots and Core Shell Systems for High Contrast Cellular/Bio Imaging

Manikanta Bayal, Neeli Chandran, Rajendra Pilankatta, and Swapna S. Nair

3.1 Introduction

Semiconductor nanostructures of particle size less than 100 nm exhibit different electronic, optical and biological properties. These structures are categorized into three—Quantum well/thin films, Quantum wire/rod/belt and Quantum dot (QD) based on the dimensions of moving particle. The qugory in which, all the particles present in the structure are confined in all the three spatial directions. This chapter mainly focuses on QD based structures and their different applications in biology. QDs with particle size less than 10 nm show drastic changes in optical absorption, exciton emission, electron–hole pair recombination etc. Therefore, the synthesis of QD is a challenging task as it should not affect the structure and its intrinsic properties. Since the intrinsic properties of QDs are determined by size, shape, surface defects, impurity and crystallinity, tailoring of them is essential for specific imaging as well as device applications. Surface to volume ratio, confinement effect of QD dependent with the size. Surface to volume ratio is one of the main factors which determines the optical, electronic and biological properties of QDs. High surface to volume ratio may enhance or reduce the whole properties of QDs such as luminescent property, quantum efficiency, optical absorption, aging effects etc. Quantum confinement is another unique property of a QD which can be observed if the size of the particle is very small and the energy level spacing is less than $k_B T$. They exhibit different emissions which is tuneable by changing the particle size (For example: CdSe QDs shows different Photoluminescence (PL) emission spectra for different sizes (Fig. 3.1) [1] or composition [2].

M. Bayal · N. Chandran · S. S. Nair (✉)
Department of Physics, Central University of Kerala, Kasaragod 671320, India

R. Pilankatta
Department of Biochemistry and Molecular Biology, Central University of Kerala, Kasaragod 671320, India

© The Author(s), under exclusive license to Springer Nature Singapore Pte Ltd. 2021
S. S. Nair and R. Philip, *Nanomaterials for Luminescent Devices, Sensors, and Bio-imaging Applications*, Progress in Optical Science and Photonics 16,
https://doi.org/10.1007/978-981-16-5367-4_3

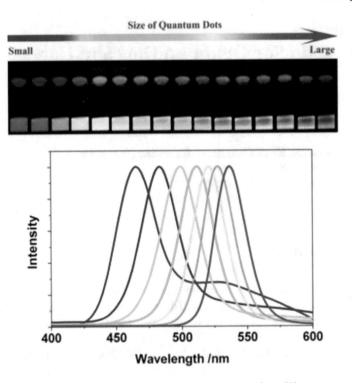

Fig. 3.1 The PL emission spectra of CdSe QDs of sizes from 1 to 10 nm [1]

Various chemical methods (co-precipitation technique, sol–gel method, hydrothermal method, chemical reduction method) were used to synthesise QDs. Chemical co-precipitation method is one of the simplest and time-consuming method for the synthesis of QDs. But this method does not assure about the stability of the QDs. There we can use other chemical methods.

3.2 Why QDs for Bioimaging?

QDs are excellent materials for bioimaging because of several reasons. (i) Its absorption and emission can be tuned with particle size (ii) The often offer more excitation window as well as narrow emission peaks (iii) it offers higher quantum yields (QYs) (iv) they often exhibit lower toxicity when compared to other NPs conjugates with organic dyes (v) They show high level of biocompatibility and may be functionalized with different bio-active agents (vi) inorganic QDs are highly photo stable under UV excitation than organic molecules [1].

While moving to the biological applications of QDs, it is essential that the QD should not affect cellular uptake and induce cell damage. Most of the QDs contain

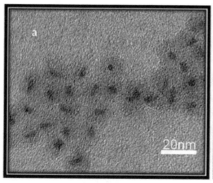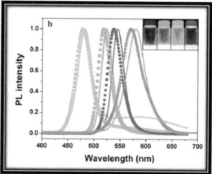

Fig. 3.2 a HRTEM image of CdSe QDs coated with silica shell, **b** field dependent emission of CdSe QDs coated with silica shell [1]

toxic ions. CdSe, CdS, CdTe etc. are few examples for toxic QDs. Therefore, it is necessary to surface modify ie, to add or cover some material like a shell over such types of QDs to reduce the toxicity. Reports show that surface modification with silica shell over the CdSe QDs largely reduce the toxicity. (Fig. 2a). This silica shell will prevent the leakage of Cd^{2+} from infra-red emitting CdTe QDs. The luminescent properties of silica shell coated CdSe QDs depends on the surface charge as well as the electric field. A field dependent emission can be seen in CdSe QDs and it is shown in Fig. 2b. Such type of emission is known as quantum confined stark effect [3]. Similar works were done by Selven et al. [4].

Traditional organic based dyes show good bioimaging [5, 6]. Therefore, the non/least-toxic dye can be used for cellular imaging application, so that it will cover the toxic QDs to protect the cellular surface and provide good images. In inorganic nanostructures, the ZnS QD is one of the less toxic material that can be explored for variety of biological applications [7]. Several works are carried using ZnS capped with CdSe QDs for reducing the toxicity [8, 9]. Here, we highlight the research works carried by several groups on cellular imaging based on QDs [10, 11].

An experiment was done by S. santra et al. on a brain tissue of a rat using TAT-conjugated CdS:Mn/ZnS QDs [12]. The results clearly showed the microscopic fluorescence images of brain tissue (Fig. 3.3). A pink and blue color fluorescence was appeared in the brain tissue due to the QD labelling and from the background without QD labelling (Fig. 3.3a–c). To get more information at the cellular level, the TAT-conjugated QDs was penetrated into the endothelial cells and it was confirmed by using transmission and fluorescence images (Fig. 3.3d and e). These results are highly beneficial for the surgical procedure. QD bioconjugation is also an important process in this study. The size of the QD is slightly equal or greater than the proteins. The conjugation of NPs surface with the molecules with high biological affinity is known as QD bioconjugation. There are two types of interactions present in the attachment of proteins and peptides with the QD. One is dative thiol-bonding between QD surface sulphur atoms and cysteine residues [13, 14]. The Weiss group

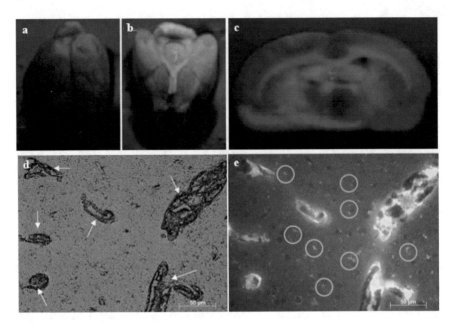

Fig. 3.3 Microscopic fluorescence images of brain tissue of rat. **a** and **b** represents the dorsal view of rat brain with TAT-conjugated QDs and **c** represents the coronal section. **d** and **e** shows the transmission and fluorescence microscope images respectively [12]. Reprinted with permission from Ref. [12]. Copyright (2005) Royal Society of Chemistry

demonstrated the same by using phytochelatin-related peptides to cap CdSe/ZnS QDs [14]. The second one is metal-affinity coordination of histidine (HIS) residues to the QD surface Zn atoms [15, 16]. Using this any peptides or proteins can be directly attached to Zn on the QD surface.

3.3 Applications of Bioimaging

Various researchers demonstrated multiple color imaging with the QDs in different cells. There are different cellular components or proteins, which show their own functions in QD labelling/imaging. Nucleus/nuclear proteins gives genetic information of the internal organelles, which is enclosed by a membrane containing pores, that mediate in and out transport. Mitochondria is another cellular component which contain its own genome, that provide essential energy delivery processes. Microtubules and actin filaments are cytoskeleton proteins, that maintains cellular structure. There is an endocytic compartment which will mediate the specific and nonspecific internalisation of extracellular molecules. Cytokeratin is a cytoskeleton protein that is over expressed and differentially stained in many skin cancers cells. Selective colouring based organelle marking using QD systems offers great advantage in this regard.

3.3.1 Cancer Cell Imaging

The good optical and luminescence properties of QDs makes them promising imaging probes for sensitive cancer imaging [17]. QDs are extremely photostable and are considered to be brighter probes as compared to the traditional organic based dye [5, 10, 18]. Their attractive optical properties promise sensitive cancer imaging properties. Continuous tracking of cell migration, differentiation and metastasis could be made possible with QDs [19]. As it is discussed earlier, the multicolor emission can be made possible due to its size tuneable optical properties. There are several reports exist which showed various strategies of delivering QDs to the cells [20, 21]. The cell uptake of QDs may depends on its size, shape or media parameters [22]. The delivery of QDs to the specific region of cell compartments such as nucleus was studied by Chen and Gerion using cell penetrating peptides [23]. Some articles successfully demonstrated the in vivo imaging for detecting tumors [20, 24, 25]. It is believed that, the QDs offer high sensitivity for tumor imaging. The image quality of tumors depends on various parameters like wavelength, light intensity, coherence, polarization, lifetime etc. Out of these, the wavelength should be very selective, for increasing the penetration capability. Some organic or inorganic dye can be used for the increase the light penetration stability, which will give images in infra-red region. For this study, the selection of dyes is also an important task. Because some dyes may lose their brightness through photo bleaching.

Variety of techniques are being developed by researchers for the entry of NPs into the cells. Direct entry of water suspended QDs are often through endocytosis process and the major drawback of this is the non-specificity as far as location inside the cells is concerned. Another technique is the direct microinjection of nanolitres of QDs inside the cells which is practically impossible for large volume labelling is concerned. Electroporation is another well sought-after technique for the entry of NPs inside the cells in which the NPs are taken inside the cells through the micro-pores induced by the electroporation. However, due to the expected cell damage, this method also needs better substitution. Recently, scientists have developed other platforms like lipid based transfecting agents for the delivery of NPs inside the cells [26]. An example is Lipofectamine 2000. Employment of cell penetrating peptides is an efficient way of delivery of NPs to the desired specific locations inside the cells. HIV derived TAT peptides is an excellent candidate for this [23]. Antibody conjugated QDs is another effective alternative which can specifically target the surface receptors thereby delivering the QDs to the desired location inside the cells [27, 28].

3.3.1.1   In Vitro and in Vivo Studies

Several invitro studies were done by various researchers using QDs and cancer cell imaging study was demonstrated in various cancer cells such as human breast cancer cellanes MCF-7, MDA-MB-435S [28] and SK-BR-3 [29], human prostate cancer [20], skin cancer (B16 melanoma) [26], human neuro blastoma (SK-N-SH) [30], Colon tumor (SW480), lung tumor (NCIH1299) and bone tumor (Saos 2) [28], etc. The experiments performed on both live and fixed human breast cancer cells (SK-BR-3) demonstrated robustness in terms of brightness and photostability of QD based labels over the traditional organic dyes [29].

There exist challenges to image cancer tissues that are deeply seeded. Application of NIR fluorescent probes are required for this purpose. This is due to the fact that most tissue chromophores, including oxyhemoglobin, deoxyhemoglobin, and melanin, absorb comparatively weakly in the NIR spectral range [31]. As an example, squamous carcinoma cells were successfully imaged in C_3H mice using bovine serum albumin (BSA)–coated CdMnTe/Hg QDs with a broad fluorescence peak in the NIR at 770 nm [32].

It has been observed that both active as well as passive targeting of QDs have been employed to deliver into tumors under in vivo conditions. Towards this surface modification of QDs was performed in two different ways; i) For passive targeting, QDs were coated with PEG and for ii) active targeting QDs were co-coated with both PEG and prostate-specific membrane antigen (PSMA) antibody. It was interesting to note that in vivo passive targeting studies of human prostate cancer growing in nude mice indicated that QDs were accumulated at tumor site by the enhanced permeability and retention whereas active targeting by PSMA antibody found to be binding to cancer-specific cell surface receptors. The use of an ABC triblock copolymer solved the problems of particle aggregation and fluorescence loss previously encountered for QDs stored in physiological buffer or injected into live animals [8, 13, 33]. QDs allow tracking of cancer metastasis in vivo by virtue of their photostability, brightness, and multicolor imaging capabilities. Most importantly, using water soluble CdSe/ZnS QDs and emission spectrum scanning multiphoton microscopy, Voura et al. (2004) were able to follow cancer metastasis processes in vivo [26]. In their study, QD-labeled B16F10 melanoma tumor cells were intravenously injected into C57BL/6 mice. The loading of QDs into tumor cells were performed by means of a lipid-based transfecting agent, lipofectamine 2000. Using a multiphoton microscopy setup, it was possible to follow tumor cells that were extravasated into lung tissue. However, QD labelling was essential to follow tumor cells. Meanwhile, tissue penetration was equally essential along with the imaging potential of QDs. For improved tissue penetration, Kim et al. (2004) reported the application of a novel CdTe/CdSe core–shell QDs (called Type II QDs), with fairly broad emission at 850 nm emission and a moderate quantum yield of ~13% for cancer surgery [34]. These QDs were highly efficient enough to map the sentinel lymph node in vivo in the mouse and pig model. Using only 5 mW cm—2 excitation, they imaged lymph nodes 1 cm deep in tissue. This work clearly demonstrated the strong possibility of using NIR QDs for real-time intra-operative surgical guidance to accurately locate and remove sentinel nodes or metastatic tumors. They have also shown that conventional NIR fluorophores, such as IRDye78-CA, photobleached rapidly in a similar excitation condition. While the future looks quite promising, there remain several encounters in the development of QD-based cancer imaging probes as its very complex. As compared to CdTe, CuInSe2 QD shows lower toxicity and better imaging at NIR (due to its band gap in the near infra-red region) [35] and hence can be exploited for the applications in which the NIR imaging potential can be used.

Silicon QDs also show great potential for biological imaging and diagnostic applications. F. Erogbogbo and team prepared highly stable aqueous suspension of Silicon QDs using phospholipid micelles for the imaging of pancreatic cancer [36]. It was found that the CdSe/CdS/ZnS QDs also shows good application in pancreatic cancer cell imaging. This QDs have better photoluminescence efficiency and stability than the CdSe/ZnS QDs. Therefore CdSe/CdS/ZnS QDs were used by researchers as optical contrast agent for imaging pancreatic cancer cells in vitro using transferrin and anti-Claudin-4 as targeting ligands [37]. Other reports discussed about the synthesis of InAs QDs with a ZnCdS shell. The results showed strong NIR emission at the wavelength range 700–900 nm and the same can be used for biological imaging applications. In this work, tumor vasculature in vivo image was demonstrated at deeper penetration depth and higher contrast than visible emitting CdSe (CdS) QDs [38].

Some literature discussed about the development of cell penetrating QDs based on the use of multivalent and endosome-disrupting (endosomolytic) surface coatings. The QDs are very smaller in size and found to be more stable in acidic environments and used for cellular uptake and imaging studies [39]. Similar works were done using graphene QDs (GQDs), which are very attractive, inexpensive, non-toxic, very biocompatible and environmentally friendly. This QDs have very smaller steric effect due to their smaller size and it can easily penetrate through smaller membranes. This unique property of GQDs makes them ideal for applications in drug delivery, in vivo and in vitro bioimaging etc .[40, 41]. Due to the excitation dependent multiple color emission of GQDs, they often show high PL quantum yield and give high contrast bioimaging with the presence of nitrogen [42]. The images obtained using different excitation wavelength is shown in Fig. 3.4. It is found that, the NIR GQDs show two

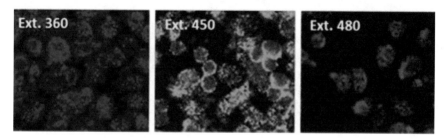

Fig. 3.4 Fluorescence images with Nitrogen doped GQDs with different excitation (360, 450 & 480 nm). Reprinted with permission from Ref. [42]. Copyright (2015) Springer Nature

photon excitations [43]. Based on this, they constructed a nanoprobe for endogenous ascorbic acid detection in living cells and its bioimaging. They also chose CoOOH nanoflakes as fluorescence quenches and it showed better imaging of ascorbic acid due to the higher penetration ability of CoOOH-NIR GQDs in tissue [43] (Fig. 3.5).

Live cell labelling is another difficult task as compared to the fixed cells and tissues. The primary study in this aspect was reported by Chan and Nie in 1998 [10]. In their study, they incubated mercaptoacetic acid coated (CdSe)ZnS QD to the cancer cells to get bright fluorescence in vivo. Another investigation was reported using dihydrolipoic acid capped CdSe/ZnS QDs with HeLa cells for long-term multicolor imaging [44]. The feasibility of aqueous CdS QDs can be exploited as an imaging tool with salmonella typhimurium cells by capping with 3 mercaptopropionic acid. In this work it is observed that, with higher pH and mercaptopropionic acid/Cd ratio of 2, the QDs exhibit stronger emission [45]. CdS QDs with DNA as a ligand shows high stability in biological media, and only minimal decreases in cell viability was

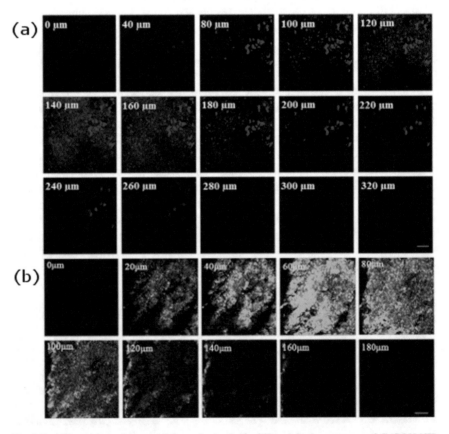

Fig. 3.5 a Fluorescence images of tissue in depth ($0-320\mu$m) in the presence of CoOOH-NIR GQDs. **b** Fluorescence images of tissue in depth ($0-180\mu$m) in the presence of CQDs. Reprinted with permission from Ref. [43]. Copyright (2017) American Chemical Society

observed when tested for toxicity in HeLa cells [46]. This material produces high wavelength luminescence which can be used for imaging application. Similar works are carried out by B. Dubertret and team [27]. In their study, they demonstrated in vitro (and in vivo imaging by encapsulating individual nanocrystals in phospholipid block-copolymer micelles. When conjugated to DNA, the nanocrystal micelles acted as in vitro fluorescent probes to hybridize to specific complementary sequences and when injected into Xenopus embryos, the nanocrystal-micelles were found to be stable, nontoxic, cell autonomous, and slow to photo bleach. The imaging properties of QD interacting with DNA was also reported by W. Zhang and his group [47]. Some other works reported the demonstration of continuous tracking of the living cells with image probing based on QDs tagged with small molecular phenylboronic acid [48]. For the first time, an image probe technique was constructed using CdSe/ZnS QDs— peptide conjugates for imaging the nuclei of living cells by F. Chen and D. Gerion in 2004 [23]. A TAT peptide conjugated QDs for imaging and tracking of living cells were reported. The study found that, the TAT-QDs strongly bind to cellular membrane structures such as filopodia and that large QD containing vesicles were released from the tips of filopodia by vesicle shedding and the results can helps us in designing the NPs for molecular bioimaging and targeted therapy [49].

3.3.1.2 Drug Delivery

Dextran is one of the best biocompatible polymers with very low toxicity. It can be used for cell detection and also can be used in nanomedicine formulations/targeted drug delivery. R Wilson et al. showed that the QDs surface modified with dextran are highly luminescent and stable and can be used for single cell imaging [50]. Based on Bi-fluorescence resonance energy transfer (Bi-FRET) mechanism a novel QD-aptamer-doxorubicin conjugate as a targeted cancer imaging, therapy and sensing system was developed by V. Bagalkot et al and team. In this mechanism, a simple multifunctional nanoparticle system can deliver doxorubicin to the targeted prostate cancer cells and sense the delivery of doxorubicin by activating the fluorescence of QD, which concurrently images the cancer cells [51]. The optical and biological properties of immunoliposome based QDs were studied for cancer diagnosis and treatment by K. C. Weng et al [52]. In this work they included tumor cell selective internalisation and anticancer drug delivery too. It was reported that the QD Carrier peptide bioconjugates exhibit no effect on cell viability and possess high stability inside the living cells. Therefore, it can be used for diagnostic and therapeutic imaging purposes for biologically active siRNA [53].

3.4 Conclusion

In semiconductor nanostructures, QD promises wide range of applications in bioimaging and targeted drug delivery. Proper surface modification and grain size tailoring can make the QD systems emit in Near UV to Near IR range and this exceptional property can be exploited for in vitro and in vivo bio imaging. The unique properties like tuneable emission, High fluorescence emission intensity, selectivity, shelf life/stability and slow photo bleaching makes the semiconductor nanoparticles as the future materials for tagless in vivo bio imaging and more and more novel engineered nano materials are being developed around the globe with lower toxicity like graphene, fluorescent carbon NPs etc. and large scope exists in surface modification of existing well sought after semiconductor materials like CdS, CdSe, ZnO, ZnS etc. for specific organelle marking inside the cells.

References

1. D. Bera, L. Qian, T.-K. Tseng, P.H. Holloway, Quantum dots and their multimodal applications: a review. Materials (Basel). **3**, 2260–2345 (2010)
2. R.E. Bailey, A.M. Smith, S. Nie, Quantum dots in biology and medicine. Phys. E Low-dimensional Syst. Nanostructures **25**, 1–12 (2004)
3. S.A. Empedocles, M.G. Bawendi, Quantum-confined stark effect in single CdSe nanocrystallite quantum dots. Science **278**, 2114–2117 (1997)
4. S.T. Selvan, T.T. Tan, J.Y. Ying, Robust, non-cytotoxic, silica-coated CdSe quantum dots with efficient photoluminescence. Adv. Mater. **17**, 1620–1625 (2005)
5. M. Bruchez, M. Moronne, P. Gin, S. Weiss, A.P. Alivisatos, Semiconductor nanocrystals as fluorescent biological labels. Science **281**, 2013–2016 (1998)
6. G.D. Luker, K.E. Luker, Optical imaging: current applications and future directions. J. Nucl. Med. **49**, 1–4 (2008)
7. R. Dungdung et al., A slow, efficient and safe nanoplatform of tailored ZnS QD-mycophenolic acid conjugates for in vitro drug delivery against dengue virus 2 genome replication. Nanoscale Adv. **2**, 5777–5789 (2020)
8. H. Mattoussi et al., Self-assembly of CdSe− ZnS quantum dot bioconjugates using an engineered recombinant protein. J. Am. Chem. Soc. **122**, 12142–12150 (2000)
9. Y. Li et al., Cellulosic micelles as nanocapsules of liposoluble CdSe/ZnS quantum dots for bioimaging. J. Mater. Chem. B **4**, 6454–6461 (2016)
10. W.C.W. Chan, S. Nie, Quantum dot bioconjugates for ultrasensitive nonisotopic detection. Science **281**, 2016–2018 (1998)
11. X. Gao, L.W.K. Chung, S. Nie, Quantum dots for in vivo molecular and cellular imaging, in *Quantum Dots* (Springer, 2007), pp. 135–145
12. S. Santra et al, Rapid and effective labeling of brain tissue using TAT-conjugated CdS: Mn/ZnS quantum dots. Chem. Commun. 3144–3146 (2005)
13. M.E. Åkerman, W.C.W. Chan, P. Laakkonen, S.N. Bhatia, E. Ruoslahti, Nanocrystal targeting in vivo. Proc. Natl. Acad. Sci. **99**, 12617–12621 (2002)
14. F. Pinaud, D. King, H.-P. Moore, S. Weiss, Bioactivation and cell targeting of semiconductor CdSe/ZnS nanocrystals with phytochelatin-related peptides. J. Am. Chem. Soc. **126**, 6115–6123 (2004)
15. J.F. Hainfeld, W. Liu, C.M.R. Halsey, P. Freimuth, R.D. Powell, Ni–NTA–gold clusters target His-tagged proteins. J. Struct. Biol. **127**, 185–198 (1999)

16. J.M. Slocik, J.T. Moore, D.W. Wright, Monoclonal antibody recognition of histidine-rich peptide encapsulated nanoclusters. Nano Lett. **2**, 169–173 (2002)
17. N. Chandran et al., Label free, nontoxic Cu-GSH NCs as a nanoplatform for cancer cell imaging and subcellular pH monitoring modulated by a specific inhibitor: bafilomycin A1. ACS Appl. Bio Mater. **3**, 1245–1257 (2020)
18. W.C.W. Chan et al., Luminescent quantum dots for multiplexed biological detection and imaging. Curr. Opin. Biotechnol. **13**, 40–46 (2002)
19. S. Santra, D. Dutta, Quantum dots for cancer imaging, in *Nanoparticles in Biomedical Imaging* (Springer, 2008),pp. 463–485
20. X. Gao, Y. Cui, R.M. Levenson, L.W.K. Chung, S. Nie, In vivo cancer targeting and imaging with semiconductor quantum dots. Nat. Biotechnol. **22**, 969–976 (2004)
21. X. Michalet et al., Quantum dots for live cells, in vivo imaging, and diagnostics. Science **307**, 538–544 (2005)
22. M. Bayal et al., Cytotoxicity of nanoparticles-Are the size and shape only matters? or the media parameters too?: a study on band engineered ZnS nanoparticles and calculations based on equivolume stress model. Nanotoxicology **13**, 1005–1020 (2019)
23. F. Chen, D. Gerion, Fluorescent CdSe/ZnS nanocrystal– peptide conjugates for long-term, nontoxic imaging and nuclear targeting in living cells. Nano Lett. **4**, 1827–1832 (2004)
24. I.L. Medintz, H.T. Uyeda, E.R. Goldman, H. Mattoussi, Quantum dot bioconjugates for imaging, labelling and sensing. Nat. Mater. **4**, 435–446 (2005)
25. C. Bremer, V. Ntziachristos, R. Weissleder, Optical-based molecular imaging: contrast agents and potential medical applications. Eur. Radiol. **13**, 231–243 (2003)
26. E.B. Voura, J.K. Jaiswal, H. Mattoussi, S.M. Simon, Tracking metastatic tumor cell extravasation with quantum dot nanocrystals and fluorescence emission-scanning microscopy. Nat. Med. **10**, 993–998 (2004)
27. B. Dubertret et al., In vivo imaging of quantum dots encapsulated in phospholipid micelles. Science **298**, 1759–1762 (2002)
28. T. Pellegrino et al., Quantum dot-based cell motility assay. Differentiation **71**, 542–548 (2003)
29. X. Wu et al., Immunofluorescent labeling of cancer marker Her2 and other cellular targets with semiconductor quantum dots. Nat. Biotechnol. **21**, 41–46 (2003)
30. J.O. Winter, T.Y. Liu, B.A. Korgel, C.E. Schmidt, Recognition molecule directed interfacing between semiconductor quantum dots and nerve cells. Adv. Mater. **13**, 1673–1677 (2001)
31. V. Ntziachristos, C. Bremer, R. Weissleder, Fluorescence imaging with near-infrared light: new technological advances that enable in vivo molecular imaging. Eur. Radiol. **13**, 195–208 (2003)
32. N.Y. Morgan et al., Real time in vivo non-invasive optical imaging using near-infrared fluorescent quantum dots1. Acad. Radiol. **12**, 313–323 (2005)
33. X. Gao, W.C.W. Chan, S. Nie, Quantum-dot nanocrystals for ultrasensitive biological labeling and multicolor optical encoding. J. Biomed. Opt. **7**, 532–537 (2002)
34. S. Kim et al., Near-infrared fluorescent type II quantum dots for sentinel lymph node mapping. Nat. Biotechnol. **22**, 93–97 (2004)
35. E. Cassette et al., Synthesis and characterization of near-infrared Cu– In– Se/ZnS core/shell quantum dots for in vivo imaging. Chem. Mater. **22**, 6117–6124 (2010)
36. F. Erogbogbo et al., Biocompatible luminescent silicon quantum dots for imaging of cancer cells. ACS Nano **2**, 873–878 (2008)
37. J. Qian et al., Imaging pancreatic cancer using surface-functionalized quantum dots. J. Phys. Chem. B **111**, 6969–6972 (2007)
38. P.M. Allen et al., InAs (ZnCdS) quantum dots optimized for biological imaging in the near-infrared. J. Am. Chem. Soc. **132**, 470–471 (2010)
39. H. Duan, S. Nie, Cell-penetrating quantum dots based on multivalent and endosome-disrupting surface coatings. J. Am. Chem. Soc. **129**, 3333–3338 (2007)
40. M.-L. Chen, Y.-J. He, X.-W. Chen, J.-H. Wang, Quantum-dot-conjugated graphene as a probe for simultaneous cancer-targeted fluorescent imaging, tracking, and monitoring drug delivery. Bioconjug. Chem. **24**, 387–397 (2013)

41. D. Iannazzo et al., Graphene quantum dots for cancer targeted drug delivery. Int. J. Pharm. **518**, 185–192 (2017)

42. D. Qu, M. Zheng, J. Li, Z. Xie, Z. Sun, Tailoring color emissions from N-doped graphene quantum dots for bioimaging applications. Light Sci. Appl. **4**, e364–e364 (2015)

43. L.-L. Feng et al., Near infrared graphene quantum dots-based two-photon nanoprobe for direct bioimaging of endogenous ascorbic acid in living cells. Anal. Chem. **89**, 4077–4084 (2017)

44. J.K. Jaiswal, H. Mattoussi, J.M. Mauro, S.M. Simon, Long-term multiple color imaging of live cells using quantum dot bioconjugates. Nat. Biotechnol. **21**, 47–51 (2003)

45. H. Li, W.Y. Shih, W.-H. Shih, Synthesis and characterization of aqueous carboxyl-capped CdS quantum dots for bioapplications. Ind. Eng. Chem. Res. **46**, 2013–2019 (2007)

46. N. Ma, J. Yang, K.M. Stewart, S.O. Kelley, DNA-passivated CdS nanocrystals: luminescence, bioimaging, and toxicity profiles. Langmuir **23**, 12783–12787 (2007)

47. W. Zhang, Y. Yao, Y. Chen, Imaging and quantifying the morphology and nanoelectrical properties of quantum dot nanoparticles interacting with DNA. J. Phys. Chem. C **115**, 599–606 (2011)

48. A. Liu et al., Quantum dots with phenylboronic acid tags for specific labeling of sialic acids on living cells. Anal. Chem. **83**, 1124–1130 (2011)

49. G. Ruan, A. Agrawal, A.I. Marcus, S. Nie, Imaging and tracking of tat peptide-conjugated quantum dots in living cells: new insights into nanoparticle uptake, intracellular transport, and vesicle shedding. J. Am. Chem. Soc. **129**, 14759–14766 (2007)

50. R. Wilson, D.G. Spiller, A. Beckett, I.A. Prior, V. Sée, Highly stable dextran-coated quantum dots for biomolecular detection and cellular imaging. Chem. Mater. **22**, 6361–6369 (2010)

51. V. Bagalkot et al., Quantum dot− aptamer conjugates for synchronous cancer imaging, therapy, and sensing of drug delivery based on bi-fluorescence resonance energy transfer. Nano Lett. **7**, 3065–3070 (2007)

52. K.C. Weng et al., Targeted tumor cell internalization and imaging of multifunctional quantum dot-conjugated immunoliposomes in vitro and in vivo. Nano Lett. **8**, 2851–2857 (2008)

53. C. Walther, K. Meyer, R. Rennert, I. Neundorf, Quantum dot− carrier peptide conjugates suitable for imaging and delivery applications. Bioconjug. Chem. **19**, 2346–2356 (2008)

Chapter 4
Tuning of Surface Plasmon Resonance (SPR) in Metallic Nanoparticles for Their Applications in SERS

Neeli Chandran, Manikanta Bayal, Rajendra Pilankatta, and Swapna S. Nair

4.1 Introduction

Noble metal nanoparticles have promising properties including large surface to volume ratio, increased surface area, structural and geometrical features and available conjugation sites on the surfaces. Surface plasmon resonance, the most exciting property exhibited by metal nanoparticles, is due to the coupling between the collective oscillation of free electrons on the surface of the nanoparticle and electromagnetic waves. More interestingly, these nanoparticles demonstrate localized effects by confining and enhancing the light in an ultrasmall volume, which is termed as the localised surface plasmon resonance. The size and shape of the nano or microparticles assume crucial role in determining their optical properties.

Surface plasmon Resonance is the phenomenon of collective oscillation of conduction electrons present on the surface of the nanoparticles when it interacts with the electromagnetic spectra. The resonance wavelength is strongly dependent on the band structure, size shape and surface properties of the NPs as well as the nature of the dispersion medium. SPR and SPR materials can offer great application potential in spectroscopic detection of single molecules through Surface Enhanced Raman Scattering (SERS) where the intensities of the Raman lines are increased by several logarithmic order due to the resonant energy transfer.

N. Chandran · M. Bayal · S. S. Nair (✉)
Department of Physics, Central University of Kerala, Periye, Kasaragod, Kerala 671320, India

R. Pilankatta
Department of Biochemistry and Molecular Biology, Central University of Kerala, Periye, Kasaragod, Kerala 671320, India

© The Author(s), under exclusive license to Springer Nature Singapore Pte Ltd. 2021
S. S. Nair and R. Philip, *Nanomaterials for Luminescent Devices, Sensors, and Bio-imaging Applications*, Progress in Optical Science and Photonics 16,
https://doi.org/10.1007/978-981-16-5367-4_4

4.2   Metal Nanostructures

Metal nanostructures are widely being used for electronic and optical applications due to their enhanced properties. Generally, metal nanoparticles have tuneable shape and size effects depending to their properties also. Noble metal nanoparticles are the most studied nanomaterials due to their wide variety of properties. In the case of metallic nanoparticles, conduction electrons are confined in all the three dimensions. The electron oscillations present at the surface of the nanoparticles induce an electric field around them, which can be much larger than the incident light. The property of surface plasmons are drastically changed when the particle size is reduced to nanoscale dimensions. The optical properties of the nanoparticles are dramatically changed by the formation of surface plasmons and its behaviours are completely changed from the bulk nature. SPR can be effectively tapped for their applications in the field of biology and medicine, sensing, energy & environmental science etc. [1].

Surface plasmons are studied scientifically by Gustav Mie through the optical properties of metal colloidal structures [2]. However, surface plasmons and associated properties in metallic nanoparticles were exploited much early even in ancient times particularly for the production of tinted glasses. Glasses used for the decors which were found in northern Italy, were stained by metallic nanoparticles including gold and copper [3]. The Lycurgus Cup is the most famous example of the application of surface plasmons in the ancient times, which shows colour variations and gradients when illuminated from inside or outside of the cup [3, 4]. Roman and Egyptian dynasties used a wide variety of metallic nanoparticles for making colorations in potteries, stained glass fabrication and colouring large church windows etc. [5–7]. After the fundamental path breaking studies of Mie, the origin of optical properties of metallic NPs are unravelled. The development of cutting-edge tools of nanotechnology, further could do manipulations in producing the nanoparticles in a controlled way by varying the parameters for particular applications in many fields.

4.2.1   Principle of SPR

As discussed above, surface plasmons are originated by the interaction between the matter and electromagnetic field which becomes predominant when a photon of incident light hits on the metal surface. At a particular angle of incidence, a portion of light couples with the electrons in the metal surface, and then moved due to this excitation [1]. The collective movements of electrons are called plasmons, who propagate parallel to the surface of metals. In detail, a metallic nanoparticles surface can be described as a lattice of ionic cores with freely moving conduction electrons. When the particle interacts with the incident light, surface electrons are exerted by a force. So, these conduction electrons are confined inside the nanoparticles. An electric dipole is generated by the accumulation of positive charges on one side and

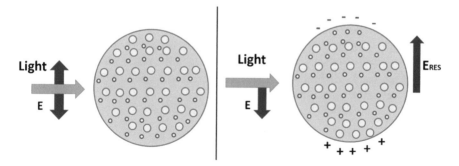

Fig. 4.1 Schematic representation of light interaction with metallic nanoparticles. Electric field of the light generates the movements of electrons and the nanoparticle's surface creating an electric dipole. This accumulation charges creates an electric field opposite to the electric field of light

the negative charges on the other side [7]. This dipole creates an electric field inside the nanoparticle which acts opposite to the incident light which acts as an exerting force on the electrons to come back to the equilibrium position. When the electron displacement is high, the electric dipole is larger and hence the restoring force too. If there is a displacement from the equilibrium position, the electrons will start to oscillate with a frequency called resonant frequency. The frequency of oscillation surface plasmons are called **plasmonic frequency**. When electric field interacts with the nanoparticle surface, an electric dipole is created on the surface. Same as above, the charge accumulation creates an electric field opposite to that of the incident light [8, 9] (Fig. 4.1).

If the plasmonic frequency of nanoparticles and the frequency of external forces are the same, the electrons will start to oscillate. When the electrons oscillations are higher, the light extinction also becomes high, and hence the absorption spectrum gives the excitation of surface plasmons. In the case of metallic nanoparticles, resonant frequency of these electrons oscillations corresponds to UV-Visible light, so the surface plasmon exhibit absorption bands in this region. In the case of gold and silver nanoparticles, surface plasmon bands are observed at 523 and 450 nm respectively. In metallic nanoparticles, there are other possible electronic transitions like inter band transition. Generally, metallic materials have continuous spectrum of states with overlapping valance and conduction bands, although, some inner electronic levels do not split enough to overlap the conduction and valance bands. So, in some cases, metallic system may exhibit inter band transitions similar to that of semiconductor materials [10]. The transition between these inner levels and conduction band originates an absorption band edge like that in semiconductors. In some metals, a weak photoluminescence is present due to the electron decay between the energy bands. For bulk materials, these optical transitions are very rare and sometimes produce with very weak signal. A feeble emission associated to the inter band transitions has been observed for bulk gold, which corresponds to the transition between 3d band and the conduction band. When the size of the particles is reduced, energy bands are not properly formed because of the reduced number of atoms, so, inter band transitions are the most distinguishable ones at this range [11].

The width and the position of surface plasmon bands depends on the size and shape of the nanoparticles. In the case of some metallic nanoparticles, the SP band and the inter band transition are overlapped. This is a critical parameter to analysing the shape of surface plasmons [12].

4.2.2 Localized Surface Plasmons Resonance

As discussed above, the SPR appears in the nanometre range is called Localised Surface Plasmon Resonance (LSPR) [13]. In the nano range, the surface plasmon behaviour exhibits in two types, localised effect and propagating plasmon. As compared to SPR, LSPR generates stronger resonance absorbance peaks, and its position is highly sensitive to the shape and size of the nanoparticles, local dielectric environment and the refractive index of the surrounding medium [14]. The SPR is applicable for the surface of bulk material also; but, the plasmons associated with the bulk material cannot be excited by ordinary wavelengths [15]. Due to the speciality of nanoparticles, the plasmons can be excited by the ordinary wavelengths because, momentum conservation is no longer required. In the case of nanoparticles, the induced electric field remains localised in a nanoscale range around the interface of nanoparticle and dielectric [15, 16]. In a LSPR, the dimension of nanoparticles is smaller than the incident wavelength. So, plasmons oscillates locally around the particles. For gold and silver nanoparticles this resonance effect is demonstrated in the visible region of the electromagnetic spectrum. Due to its sensitive nature, metallic nanoparticles can be employed as a potential plasmonic biosensor and chemical sensors [17] (Fig. 4.2).

In the other case, the induced surface charge oscillations on the nanoparticles couple with the electromagnetic waves and starts to propagate at the metal-dielectric interface. These are generally called **surface plasmon polaritons (SPP)** [18]. In SPP,

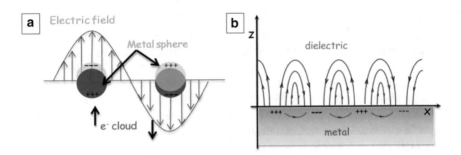

Fig. 4.2 Schematic representation of SPR [17]. Reproduced with permission. Copyright © 2016, Royal Society of Chemistry

plasmons propagate along the metal- dielectric interface in the X and Y directions. This is precisely confined in two-dimensional space. Although, the propagation of SPP have damping, this is the key issue when it is used as plasmonic sensors and other optical applications. Propagating electromagnetic waves which are transverse in nature are called **Surface Plasmon Waves (SPW)**. They can be of radiative or nonradiative nature [18].

Due to the special characteristic of nanoparticles, the ordinary visible light is enough to initiate the plasmonic oscillation. In nanoparticles, due to the coupling of bulk and surface plasmons, charge densities are spread all over the surfaces. In the case of spherical particles, dipolar plasmons are produced [19]. If the particles are in different shape, multipolar plasmons can exist for individual nanoparticles. The ultimate difference between dipolar plasmon and multipolar plasmon is the shape of the surface charge distribution [20, 21]. The tunability of LSPR properties in wide range of wavelengths depends on the structure of nanoparticles. In the case of nano-shells and nanorods, LSPR can be tuned into near-infrared region, which is useful in in-vivo imaging techniques and cancer therapy [22]. Gold, Silver and Copper NPs are widely used for biomedical applications because they possess much better SPR and other plasmonic properties [14]. In the presence of these metals, the core-shell system and nanocomposite systems have better performance in therapeutic and diagnostic applications [23].

4.2.3 *Metal Nanoparticles and the Theory of Mie*

The colour change observed in colloidal gold nanoparticles with particle size inspired Gustave Mie to study about the light extinction in small particles [2]. He applied the general theory of light extinction to the nanoparticles. Mie theory is built upon the basis of Maxwells equation with suitable boundary conditions in multipole expansions to determine the electrodynamic calculations of interaction between the light and the spherical metallic nanoparticles. Theory gives the frequency dependent response of spherically shaped particles to the interacting electric field in terms of real and imaginary part of the dielectric function. Surface plasmonic band width, position and intensity in the extinction spectra depends on the dielectric function and the size of the particles. The shape and composition are the other factors which can influence the characteristics of surface plasmons in metal nanoclusters. Red shift of surface plasmonic peak occurs when size of the particles increases. Most of the colloidal synthesis methods give the particles that are spherical in nature and Mie theory is mainly applicable in the case of spherical particles. Some assumptions are used in the **Mie theory**. The surrounding medium and the particle are homogenous and their bulk optical parameters are used to describe the system. This is the first assumption of Mie theory [24, 25]. The electron density is determined at the boundary of the particle; but there is a sharp discontinuity in spherical particles at radius R. We considered that the size of the particle with radius R is much smaller than the wavelength of light ($2R \ll \lambda$). This condition is used to relate the dielectric constant and dipole plasmon frequency [15, 26]. In this approximation, electric field is taken as constant and the interaction is controlled by the principles of electrostatic theories. This is a quasistatic approximation because the phase shifts of electrodynamic field over the particle diameter are negligible. If the particle size is greater than the wavelength ($2R > \lambda$), the optical light scattering becomes effective and the quasistatic approximation does not exist [13, 24].

There are some limitations in Mie approximation, because the polydispersity and matrix effect are not considered. Furthermore, it is assumed that the particles are situated as individuals and they are non-interacting each other. So, the interaction of the electric field generated around the particles by the excitation of plasmonic resonance with the surrounding particles is not taken into account [1, 2]. Practically, the coupling effect between the plasmonic resonance of neighbour particles are affected on the extinction spectra and optical properties [13]. Generally, Mie theory can describe the optical extinction spectra of particle which are small in size as compared to the wavelength and also with the lower concentration of particles [2, 27, 28]. So, Mie theory is directly applicable to the spherical particles with any size. The limitations of the theory are that the dielectric constant of the particles is changed when the size of the particles reduced from bulk to nano range [29, 30].

4.2.4 Tunability of SPR

SPR bands of the metallic nanoparticles are tuneable from visible to NIR range by changing the shape, size, growth mechanisms and nature of the surrounding medium including pH, refractive index etc. In the case of gold nanoparticles, effects of these parameters are widely studied and they are well employed for different applications. These factors are theoretically explained by the Mie theory.

4.2.4.1 Size Effect

SPR of the smaller particles with size is less than 20 nm can be explained by the following equation [2, 31].

$$C_{ext} = \frac{24\pi^2 R^3 \varepsilon_m^{3/2}}{\lambda} \frac{\varepsilon_i}{(\varepsilon_r + 2\varepsilon_m)^2 + \varepsilon_i^2} \tag{4.1}$$

Λ is the incident light and C_{ext} is the extinction cross section, it is related to the extinction coefficient. ε is the dielectric constant of the metal, which is a complex system given by $\varepsilon = \varepsilon_r(\omega) + i\varepsilon_i(\omega)$, where, $\varepsilon_r(\omega)$ is the real part and $\varepsilon_i(\omega)$ is the imaginary part of the dielectric constant. ε_m is the dielectric constant of the surrounding medium which is frequency independent and function of refractive index of the medium by the equation $\varepsilon_m = n_m^2$, n_m is the refractive index of the medium [32–34].

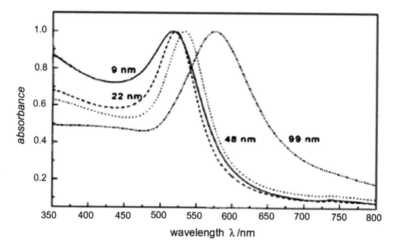

Fig. 4.3 Absorption spectra of plasmonic gold nanoparticles with different sizes [35]. Reprinted with permission [36]. Copyright© 1999, American Chemical Society

The SPR position and bandwidth of metal nanoparticles are determined by the real and imaginary part of the dielectric constants respectively. The SPR condition is fulfilled at $\varepsilon_r = -2\,\varepsilon_m$, if ε_i is small or weakly dependent to angular frequency ω [33–35].

In the case of gold nanoparticles, SPR band is often observed at around 520 nm which is affected by the particle size [36]. When particle size increases, the SPR band is red-shifted and the intensity increases, as shown in Fig. 4.3. If the size of the nanoparticles are smaller than 10 nm, the SPR band of Au NPs are largely damped because the rate of electron-surface collisions are increased as compared to the larger particles, which leads to phase changes.

Figure 4.3 shows the surface plasmon peaks of gold nanoparticles with 22, 48 and 99 nm size, which is prepared by the citrate reduction method in an aqueous solution. The molar extinction coefficient of the particles increases linearly with the increasing volume of the nanoparticles [36]. As discussed above, it can be observed that the SPR bands is red-shifted with increasing particle sizes. The SPR band width also increases above 20 nm size range. Nevertheless, the dipole resonance approximation is no longer valid for larger nanoparticles with size greater than 20 nm. The particles become larger; higher order modes are more dominant that the light cannot polarize the particle homogenously. These higher order modes occur at lower energies, so the plasmonic bands are red shifted (moves to the higher wavelength) when the particle size increases [33, 36].

The increase of plasmonic bandwidth also follows Mie theory. Optical properties of the metal nanoparticles directly depend on the size of them, which is considered as the extrinsic size effect. Band width is related to the coherent electron oscillations. The wide band width corresponds to rapid loss electron motion. The SPR bandwidth is used to compute the electron dephasing time, which strongly recommends that the main relaxation process includes electron-electron collisions. When the particles size is greater than 100 nm, the broadening of bands are obvious due to the contribution of higher order electron oscillations [32, 36].

Large number of synthesis methods are used to produce nanostructured metal particles. Synthesis of nanomaterials are broadly classified into two categories—"top down" and "bottom up" methods. Top-down method is breaking down the materials into smaller sized particles using chemical reactions which includes chemical reduction, precipitation techniques. Bottom-up approach is the building up of material structures from atom by atom using different physical and chemical methods. Mechanical grinding, ultrasonication, chemical vapour deposition, physical vapour deposition and laser ablation belongs to this category. Usually, colloidal metal nanostructures are synthesized by chemical reduction methods using suitable reducing agents. Colloidal gold nanostructures are popular, and it is synthesized by Turkevich method and citrate reduction method. Similarly, Silver and Copper nanoparticle can be synthesized by chemical reduction, hydrothermal and precipitation techniques.

4.2.4.2   Shape Effect

The surface plasmonic band of the nanoparticles are strongly depends upon the shape and structure of particles. Different synthesis methods can achieve different sized particles with various morphologies [37]. Wet chemical synthesis methods are the generally used methods to synthesize colloidal metallic nanoparticles which exhibits tuneable SPR properties [38]. The colour of the colloidal system also changed with the change in size and shape. Metallic gold nanoparticles are widely studied material with changing the parameters [37, 38].

As discussed above, spherical Au NPs have SPR band around 520 nm, which synthesized by the chemical reduction method with sodium borohydride. The citrated mediated synthesis offers elongated particles. The surface modifying agents governed the shape and morphology of the particles. CTAB and TOAB etc. are provided the directional group of the particles, thus it produces elongated Au NPs. CTAB assisted seed mediated synthesis of gold nanorods (Au NRs) are most popular method, which is achieved by the addition of gold seeds into the mixture of reducing agent and CTAB [38, 39]. Then the solution kept undisturbed overnight to allow the growth of Au NRs. Plasmonic resonance bands of Au NRs are tuned over 650–1500 nm by varying the aspects ratio or length of NRs [35, 39].

Figure 4A shows the SEM images of Au NRs with different aspects ration. The colour changes and the gradual shift of plasmonic bands with respect to increasing the aspects ratio also shown in Fig. 4.4. Spherical plasmonic particle contains single SPR band in the absorption spectrum. But the Au NRs contains two plasmonic bands are exhibited in the absorption spectrum, longitudinal plasmon band (LPB) and transverse plasmon band (TPB) [37, 39]. These LPB and TPB corresponds to the electron oscillations along the long and short axis of the Au NRs, respectively. The transverse band is independent to the changes in the size and environmental changes of the NRs. But, the longitudinal plasmonic band red-shifted by the increase in aspects

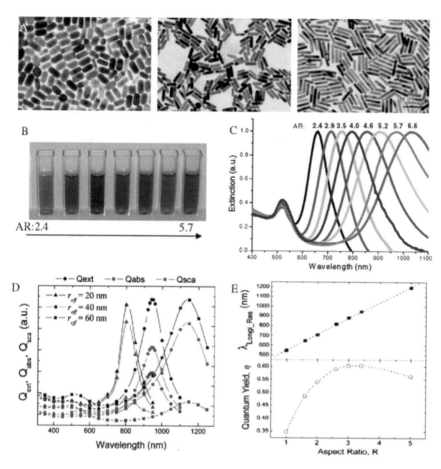

Fig. 4.4 The extinction spectra of gold nanorods with increasing aspects ratio. The colour change of the colloidal Au NRs [35]. Reproduced with permission from [35] and [42]

ratio of NRs [39]. This peak is very sensitive to the refractive index changes in the environment [26]. So, Au NRs system is employed for the biosensing applications [40–45].

4.2.5 Surface Enhanced Raman Scattering (SERS)

Surface Enhanced Raman Spectroscopy or Surface Enhanced Raman Scattering is a surface sensitive used for the detection of molecules that depends on the enhancement of Raman scattering by the molecules adsorbed on the SERS active surfaces like metallic gold and silver nanoparticles etc. [46, 47] SERS was first observed by Fleischmann et al. [48], who reported large Raman peaks from pyridine conjugated

silver electrode in 1974. Then, Moskovits suggested that the strong signals were originated by the surface property of metallic nanostructures, that is the optical excitation
of collective oscillations of electrons on the surface [49]. The SERS enhancement
is related to both the electromagnetic and chemical effects. SERS effects are opens
lots of advantages to detect and characterize the molecules, enhance the recognition
capabilities etc. Many forms of substrates including gold, silver and copper etc. have
been used to exploring the techniques since the discovery of SERS. Morphologies
and patterning of metal substrates are changed to make the maximally enhancing
SERS active substrates [50–52].

4.2.5.1 Principle of SERS

Generally, when the photons interacted with the matter, elastic or inelastic scattering
occurs. In elastic scattering, both incident and scattered photons have same energy,
this is known as Rayleigh scattering. In second case, scattered photons have lower
or higher energy than that of the incident photons, because they lose or gain energy
by interacting with matter [50]. These are called stokes and anti-stokes Raman scattering. Stokes and anti-stokes lines are corresponds to the interaction between photos
and a molecule in its ground and first vibrational excited states. So, the scattered
photons contain information about the vibrational modes of the molecule which
used for investigation. The finger print region of the Raman spectrum is material
specific and they give the important information about the structural properties of
the materials including the vibrational modes in the molecules [53, 54]. The power of
the Raman signal (P_{Raman}) is directly depending on the differential cross section. The
cross section of anti-stokes lines are much smaller than the stokes lines, therefore,
only Stokes Raman bands are normally detected [55, 56] (Fig. 4.5).

If the molecules are attached to the suitable metallic nanostructured which have
SPR properties, the low efficiency of Raman scattering can be improved. This the
case of Surface Enhanced Raman Scattering (SERS). The cross section of Raman
scattering is lower than Raleigh scattering and fluorescence. As compared to this,
the SERS cross section is much higher than the cross sections of above discussed
processes. SERS signals can be expressed in terms of power of Raman signal [55,

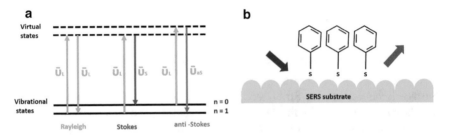

Fig. 4.5 Schematic representation of principle of SERS

57].

$$P_{SERS} = G_{SERS}.P_{Raman} \qquad (4.2)$$

G_{SERS} is the SERS enhancement factor. In the presence of nanostructures, there is considerable increase in the signal.

Electromagnetic and chemical enhancements are the two main factors are helps to enhance the Raman signal in the metal substrate in the vicinity of scattering molecules [27, 58]. Electromagnetic enhancement to SERS is owing to the surface plasmons which originates strong localised field, Raman scattering occurs from the molecules with the incident field [59]. Electromagnetic enhancement is occurred by the excitation of surface plasmons on the surface of the metallic nanostructures. When a molecule which undergo Raman scattering, treated with the intense electromagnetic field produced on the metal surface, thus the stronger polarisation of the molecule leads to produce higher induced dipole moment [60]. Electromagnetic enhancement is the major component in the SERS enhancement mechanism. Electromagnetic enhancement includes two different steps, local field enhancement and radiation enhancement. The power of the Raman signal radiated by the molecule depends upon the surrounding medium in which it is embedded. If the molecule situated in an inhomogeneous environment, radiate different amount of energy compared to the same molecule in the homogenous environment [27, 55].

Chemical enhancement is another component in SERS, which is originated due to the charge transfer between the adsorbed molecules and metal nanostructures. The involvement of chemical enhancement is less and it is depending on the chemical structure of the molecule. The polarizability of the molecules modified when molecules interacted with the surface of SERS substrate; with the help of suitable receptors to enhance the affinity towards the specific molecules. This involves the charge transfer transition between the molecule and metal complex. So, strong improvement of the certain Raman bands can be observed, due to the resonance Raman Effect. The occurrence of a CT state requires the proximity in energy of the Fermi level of the metal with the highest occupied (HOMO) or with the lowest unoccupied molecular orbitals (LUMO) of the molecule: this condition is not so uncommon; in fact, the Fermi level of metals lies in between the HOMO and LUMO of many organic molecules [13].

Electromagnetic enhancement is the main contribution on SERS, so researchers focus on the novel plasmonic structures by varying size, composition and dimension to obtain high enhancement factors. As discussed above, size and shape of the nanoparticles are the two important factors influencing the plasmonic properties of metallic structures. Resonance frequency of the surface plasmons can be tuned when the physical properties of the nanoparticles are changed. A single spherical metallic nanoparticle and illuminated particles also exhibit the same phenomenon [56, 61]. It can be extended to an array of particles with different sizes and shapes. The enhancement of SERS and excited surface plasmons are correlated which includes the tuning of surface plasmon resonance with size, shape and dielectric constants. In core-shell and colloidal gold nanoparticles the influence of size on SERS enhancement have been reported [62]. From the studies, the maximum enhancement is observed when the following Eq. 4.2 is satisfied.

$$\lambda_{sp} = \frac{\lambda_{exc} + \lambda_{RS}}{2} \tag{4.3}$$

where, λ_{exc} is the incident wavelength. λ_{sp} and λ_{RS} are the surface plasmon wavelength and Raman scattering wavelength respectively. Higher SERS enhancement factors are reached when the Surface plasmon wavelength λ_{sp} is located in between λ_{exc} and λ_{RS} [63–66].

In this section, we discuss about the influence of morphology, size and composition of nanoparticles and surface plasmon wavelength on SERS. Y. Fleger et al. studied the optimal conditions of Au, Ag and alloyed Au–Ag nanoparticle by varying the size and compositions for fabricating SERS substrates with maximum enhancement factors. SERS enhancement factor directly proportional to the fourth power of the intensity of electromagnetic field generated on the plasmonic nanostructure substrates. Enhancing electric field intensity on the surface of engineered SERS substrate is one of the main tasks. Electrodynamic calculations like discrete dipole approximation (DDA) [67] and finite difference time domain [68] methods are generally used for estimating the resonant frequency and field intensity of surface plasmons.

We discussed already about the influence of physical properties on their plasmonic properties and their SERS enhancement factor. The distance between molecule and substrate, type of structures and aggregation of particles are the other parameters influencing SERS. Molecules under investigation must be covalently bound to the substrate (or in close vicinity) in the purpose of obtaining a significant Raman signal. When the molecules and substrates are close to each other, large enhancement is obtained [69, 70]. Two types of SERS substrates are generally used in SERS experiments. Colloidal nanoparticles suspensions and solid substrates. Colloidal nanoparticle suspension is commonly used due to their simple synthesis methods and relatively better enhancement factors [58]. The distance between molecules and nanostructures SERS substrate is influenced on the type of interactions. Charge transfer mechanisms in the molecules are influenced on the Raman signals. The surface charge of colloidal nanoparticles and the analyte molecules must be considered. The aggregation state, pH of the solution and the zeta potential are depending on the surface charge of the nanoparticles and SERS activity becomes stronger when the detected molecule possess opposite charge of the nanoparticles [58].

4.2.5.2 Materials for SERS Substrates

In addition to Au and Ag, metals including Cu also exhibit plasmonic properties in the visible region, thus it can be used as SERS substrate [71]. Al and Cu based nanostructures are low-cost materials than other metals but their oxidation prone surfaces and low enhancement factors are the disadvantages. In Ag (300–1200), tunability range of SPR is wider than Au (500–1200 nm) [61, 72, 73]. Due to the high enhancement factor and wider tunability range Ag is an excellent choice for SERS substrates [58].

The physical characteristics of specific materials used for SERS experiments including size, shape, structure, composition and type of the material (2D or 3D) are also important. Figure 4.6 shows the scanning Electron Microscopy images of engineered SERS substrates fabricated by different methods. Generally, three types of SERS substrates are used for measurements; colloidal, solid and flexible substrates. The details regarding the substrates are discussed in the following sections.

Colloidal Structures

Due to the simple methods of synthesis and their size and shape controllable nature, colloidal metallic nanoparticles are directly used as the SERS substrates. Colloids exists in solution mainly water, because they stabilised by steric or columbic repulsions among particles [83, 84]. So, colloidal nanoparticle requires a surfactant or stabilising agent, which protect the surface of them and prevents from aggregation.

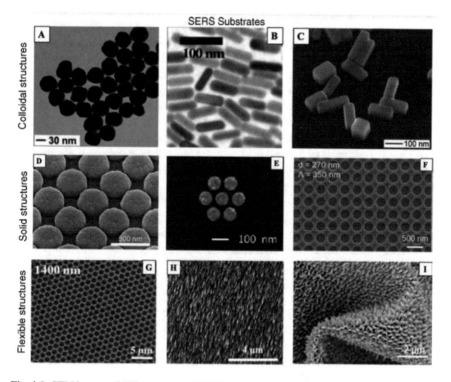

Fig. 4.6 SEM images of different types of SERS substrates. **a** Spherical gold nanoparticles, **b** gold nanorods, **c** silver nano bar, **d** silver plasmonic nano dome array, **e** gold nanocluster, **f** gold nanoholes, **g** silver nanovoids, **h** silver nanocolumnar film, and **i** silver nano-pillars [58]. Reproduced with permission from Refs. [76–84]

In some cases, single chemical compound can be act as both reducing agent and stabilising agent.

Gold (Au) nanoparticles are synthesized by citrate reduction or Turkevich method which consists of different steps; the reduction of Au ions into chargeless seeds, formation into cluster of these seeds and then, the coalescence of clusters forms nano sized particles. Colloidal Ag solution also produced through this citrate reduction methods. Hybrid materials or nanocomposites that are combination of noble metallic nanoparticle with other materials also obtained for multi-functional structures and morphologies. When increasing the particle size, the SERS activity of spherical Au NPs can be improved. Nanorods, nanoplates, nanobars, nanowires and nanorods are exhibit different SERS properties. Sometimes, hollow structures show better enhancement in SERS. Bimetallic nanoparticles can be prepared by different methods especially though wet chemical route. It is chemical reduction two metal alloys with the help of suitable surfactants. It can also be synthesized heterogeneously with the reduction of two metal ions are called core shell nanoparticles. Different types of metal alloys and core shell nanostructures are achievable like Au coated Ag NPs (Ag@Au), Au coated Cu NPs (Cu@Au) and Ag coated Au NPs (Au@Ag) etc. [84–86]. The composition and ratio of materials in colloidal systems are strongly depends on the SERS enhancement. The bimetallic structures have better enhancement than their single metallic counter parts. In core-shell systems, the thickness of shell and type of the metal core are the critical factors for SERS activity [87].

In physisorption, it depends on the charge of the analyte relative to the charge of the metal nanoparticles [88]. Depending on the synthesis methods, it can be achieve positively or negatively charged nanoparticles. To keeping the colloidal system stable and homogenous, the particle must carry charges. When the analyte and nanoparticle carry opposite charges, analytes interact with the nanoparticles with attractive forces [89]. The zeta potential of the colloidal solution indicates the degree of electrostatic repulsion between the charged particles. If the charges of nanoparticles and analytes are opposite, zeta potential reduces and thus induce the aggregation of particles. Aggregation of nanoparticles enhanced the SERS activity [90]. The surface plasmonic peak of aggregated particles also moves towards longer wavelengths with wider spectrum. Physisorption of analyte into the molecules includes electrostatic interaction, Vander Waals force interaction and hydrogen bonding etc. To improving the binding affinity of analyte, the metallic surfaces improves with specific surfactants like polymers, organic compounds, electrostatic ligands and peptides etc. They help to immobilize the analytes via physisorption or chemisorption [91–93].

In chemisorption, analytes are covalently bound to the surfactants or directly to the metal nanoparticles, and improve the binding affinity and molecular selectivity (Fig. 4.7).

Surface functionalising agents bearing functional groups such as amine(-NH2), thiol(-SH) and carboxyl(-COOH) groups can be easily bind to the metallic nanoparticles [93, 94]. Metal-thiol linkages are very strong; thus, thiolate gold nanoparticles colloids are used for SERS sensing and other several applications. Similarly, analytes containing these functional groups can be bind to the suitable receptors (capping agents) or they can bind to the metallic core by replacing the capping agents [89, 93].

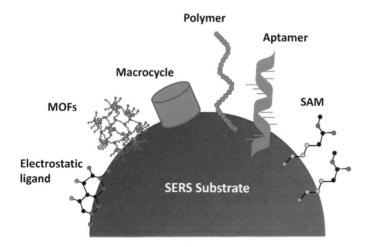

Fig. 4.7 Scheme of different surface functionalising agents attached with the nanoparticles surface through physisorption or chemisorption

Carefully surface functionalised and well stabilised colloidal nanoparticles system is necessary for sensing research. As discussed above, functionalised nanoparticle surfaces with appropriate receptors has detected guest analytes in the sample solution through biochemical and physiochemical interactions.

Stability of the colloidal substrates are an important factor. Colloidal solution contains three-dimensional detection volume as compared to the area of two-dimensional solid substrates. So, average number hotspots also much higher in colloidal solution than solid substrates. Due to this reason, colloidal substrate systems show better reproducibility. The surface chemistry, surfactants used for the synthesis of colloidal nanoparticle and the external environment ensure the stability of the system. If the concentration of targeted molecule is very low, ultra-highly sensitive detection is needed and the SERS signal may fluctuate constantly. Although, colloidal substrates have better performance than solid substrates. The zeta potential measurements are essential in colloidal solution to measuring the state of aggregation.

Solid Substrates

Solid substrates have some advantages over colloidal substates. They are more convenient to use than colloidal substrates. A good substrate must offer the following features; reproducibility and stability in SERS measurements, high average enhancement in SERS signal, large surface area, ease of fabrication and cleanliness on the surface of substrate etc.

In device fabrication, biosensors as point of care diagnostic tools, small sized 2D or 3-D substrates are more effective. 2 and 3-dimensional engineered plasmonic structures are widely used in SERS experiments. 2D substrates are fabricated with 2D sheets or films, they employ the propagating surface plasmons. They produce unique plasmonic properties because, free electrons are confined in the metal surface [95].

Two-dimensional solid substrates are array of patterned substrates with particular nanostructures. Thin film plasmonic nanostructures are by arranging the assembly of nanoparticles or thin atomic layer 2-dimensionally through chemical vapour deposition or other film fabrication techniques [96, 97]. The orientation, size, shape and interparticle distance in the nanoparticle array are key factors to determine the SERS performance. Typically, plasmonic biosensors are fabricated on the gold and silver films. Three dimensional plasmonic nanostructures have been successfully fabricated in different shapes including nanovoids, nanoholes, nanoclusters and nanoarrays [97, 98]. They can be fabricated by electron beam lithography, ion beam lithography and electrochemical deposition techniques.

In electrochemical deposition and electrochemical roughening method contains a three-electrode system which contains a reference electrode, working electrode and counter electrode, depending the process is carried out with the potentiostatic or galvanostatic condition. Working electrode of the system must be a good conductor or a foreign substrate like glassy carbon or ITO in electrochemical deposition method [99, 100]. Generally, gold, silver and copper are deposited through this method. The concentration of salt, organic ligands and concentration of electrolyte can produce significant modifications in the shape and dimension of deposited films. Electrodeposition gives large area substrates with reproducibility.

Another method to fabricate SERS substrate is the nanoparticle are adsorbed on the solid surface with the help of organic compounds containing functional groups. Natan et al. achieved a method to fabricating an array of spherical metallic nanoparticles on substrate. If a glass substrate is immersed on a (3-Aminopropyl) trimethoxy silane (APTMS) solution, the methoxy group was replaced with hydroxyl groups and this attached on the glass substrate by leaving the terminal $-NH_2$ groups exposed. When this substrate is immersed in gold or silver nanoparticle solution, that assembled on the surface and forming a layer. Further aggregation of nanoparticles on the surface is avoid by drop casing the nanoparticle solution on the modified substrate in controllable manner.

3 dimensional structures contain, large number of hotspots, which is the region intense localisation of surface plasmons and the enhancement of SERS in this signal. Engineered 3D structures with nanoparticles and nanorods are designed to generate more hotspots, the signals also have significant enhancement [101]. 3D plasmonic biosensors have better sensitivity due to the presence of increased number of hotspots. 3D nanocups and nano bowls are fabricated to get high density of hotspots. Deposition of metallic particles like gold and silver into the 3D substrates like paper, silicon wafer and glass substrates also possible to make efficient SERS substrates [89, 102, 103].

The reproducibility of the SERS measurements depends on the structural instability of the hotspots presents in the SERS substrates. When they undergo melting and laser illumination, the hotspots may face some changes by changing the size, shape and interparticle distance of the nanoparticles. In some cases, the nanostructures on the surface easily prone to surface oxidation, thereby decreasing the surface energy by changing shapes of nanoparticles to exist in more stable structures (spherical shape). This effect on the enhancement factor of SERS signal [89].

4.2.6 Applications of SERS

4.2.6.1 Biosensing

Most studied and widely used application of plasmonic nanoparticles are plasmonic biosensor which includes the SPR based and SERS biosensors.

Mainly these two techniques are used for the effective detection of molecules, proteins, chemical component, or biomarkers of several life-threatening diseases etc. Plasmonic nanoparticles act as the transducers which converts very small changes in the environment into spectral shift in the extinction spectra. From the previous section, we understood that the sensitivity of plasmonic peaks to the size, shape, pH and refractive index of the environment etc. Mainly, SERS and SPR sensors are used to the sensing of biological components and biomarkers. Biomolecules have higher refractive index than the solution. When the targeted biomolecules attached with plasmonic nanoparticle surface, the refractive index again increases and the SPR band shifted towards longer wavelength. In SERS, when the targeted analytes conjugated to the receptors in the SERS substrates, the Raman signals are altered in accordance with the concentration of analytes. Direct SERS based determination of viruses is possible in combination with oligonucleotides as the recognition elements. Multi-layered nanorod arrays of Au/Ag have been employed for the detection of H1N1, H2N2 and H3N3 influenza A virus strains with concentration of 10^6 pfu/mL. Recently, Caterina Credi et al. described the method of SERS sensor with fibre optics technology for the specific detection of amyloid-β (Aβ) which is neurotoxic biomarkers in liquid samples. SERS and SPR based biosensors are the most effective sensing methods to the early detection of ultra-small concentration of biomarkers in the human body. SERS based biosensing applications are tabulated in Table 4.1. The mechanism of biosensing, sensor device fabrication and types of biosensors are described in the next chapter in this book series.

4.2.6.2 Bioimaging and Cancer Therapy

In addition to the biosensing applications, tuneable plasmonic nanostructure offers potential nonradiative and imaging properties. If controlling the synthesis parameter of gold nanoparticles, SPR can be tuned to NIR region with particular wavelength. 50 nm edge sized Au nanocages exhibits SPR peak around 800 nm. These nanocages dominated by absorption and they are suitable for photothermal therapy applications [111, 112]. Au and Ag NPs have wide range of SPR tuneability into NIR region, so they can be employed for therapeutical applications in the biomedical field. El Sayed et al. reported that Au NPs can convert the absorbed light into heat through a series of nonradiative process [113]. The energy conversion process starts through electron-electron collisions, this leads to the formation of hot electrons with temperature 1000 K. Then the electron transfers the energy to phonons present in the lattice through electron-phonon interactions. The electron phonon relaxation process are

Table 4.1 SERS based biosensing applications

Analyte	Functionalization	Detection method	Material/substrate	Sensitivity	Ref.
Escherichia coli O157:H7	Au nanorods functionalized with 4-ATP and ATT and antibodies	SERS	Nano-DEP microfluidic device Au nanorods	10 colony-forming unit per mL	[104]
Mercury ion Hg^{2+} in water	Assembling oligonucleotide probes	SERS	Silver nanorod array by OAD technique. Oligonucleotide probes are assembled through Ag–S covalent bond	0.16 pM	[105]
R6G	None	SERS	Paper with different porosities decorated with Ag nanostars	11.4 ± 0.2 pg	[106]
Melamine in milk	None	SERS	Silver dendrite	7.9×10^{-7} M	[107]
Methylene-blue in fish graphene	None	SERS	Ag nanoflowers sandwiched between PMMA and monolayer	10^{-13} M	[108]
TNT explosive	PABT (Raman signal) modification	SERS	AgNPs modified by PABT decorating silicon wafer	$\sim$1 pM	[109]
Bromate	R6G in water used as a reference sample with or without bromate	SERS	Ag film made by electrostatic immobilization of AgNPs on glass slide	0.01 μg L^{-1}	[110]

shape and size independent. Longitudinal and transverse plasmonic peaks also does not depend on this process. Plasmonic nanoparticles exhibits resonance scattering and they offer bright colours in the dark field microscopy, they can be used as an efficient candidate for bioimaging applications [114]. Size effect and shape effect of plasmonic particles and their colour changes also described in the previous section. A silver nanosphere with 80 nm size scatters blue light (445 nm) with scattering cross section 3×10^4 [112], which is much greater than the fluorescence cross section by the scattering of fluorescein molecules. SPR imaging of cancer cells also carried out by the functionalised Au NPs conjugated with the antibodies to the cancer bio markers. In addition to the higher scattering cross section, these plasmonic nanoparticles have brilliant photostability as compared to the organic dyes, makes them superior candidate for cellular imaging applications. The SERS technique can be used as the efficient spectroscopic imaging probe. Generally, Raman tag is used the organic fluorescent dye molecules which has high Raman scattering cross sections. This fluorescence is quenched when they are attached to the metallic nanoparticles surface and Raman signals can be detected. Plasmonic nanoparticles are conjugated to both the Raman tags and cancer targeting ligands in cancer diagnosis [43, 115].

Dark field microscopic imaging of cancer cells with functionalised Au NPs and Ag NPs are described by prasad et al., they emitting bright blue and green colours respectively. Gold nanoparticles are functionalised with suitable bioreceptors which is conjugated to the specific antibodies to the cancer cell [116]. Plasmonic particles gives specific interaction with the cancer cells with bright imaging properties [43, 115]. So, this can be used to fabricate high performance plasmonic probes for the specific detection and imaging of live cancer cells. The proper surface modification of plasmonic particles is the key factor in this work. PEG modified or thiolated Au NPs are easily conjugated with the antibodies to the cancer cells [117]. Recently, Myeong Soo Kim et al. described a novel designing of magnetic-plasmonic nanoparticles assembly for targeted cancer cell imaging. In this work, gold seeds are conjugated with magnetic template through chemical binding is containing abundant hotspots. Thus, it helps to improve the SERS enhancement factor and the system is employed for targeted cancer cell imaging applications. Nuclear targeted Silver nanoparticles (NLS/RGD-Ag-NPs) used for real time plasmonic imaging of HSC-3 cells during and after induction of apoptosis [118]. Plasmonic scattering property of particles widely used in disease detection and treatment in recent times.

4.2.6.3 Drug Delivery and Photothermal Therapy.

Size and shape tuneability, long term stability, large surface to volume ration, less toxicity and easy functionalisation etc. are the specialities of Au NPs, which recently considered for targeted drug delivery applications [119, 120]. Modification of the surface of plasmonic nanoparticles are governed the targeted drug delivery [121, 122]. The controlled release of therapeutic agents could be monitored by internal or external factors. In cancer therapy, plasmonic nanoparticles used as the platform for developing multifunctional tumour-targeted drug delivery vectors [123]. The available hyperthermia treatments possess low special selectivity in heating tumour cells. So, it results the damages of the surrounding healthy cells. To improve the spatial selectivity, labelling the tumour cells with plasmonic nanoparticles is an efficient method. Photodynamic therapy [124] and plasmonic photothermal therapy (PPTT) are the technologies improved with the help of conjugated nanoparticles. By providing the nanoparticles to laser irradiation near their SPR absorption band, it is possible to heating the labelled cancer cells specifically without damaging the surrounding healthy cells. By varying the size, shape and structure of nanoparticles to achieve the absorption and scattering efficiencies and also possible the tuning of SPR bands into the therapeutical optical window, which is in the 750–1100 nm range [125].

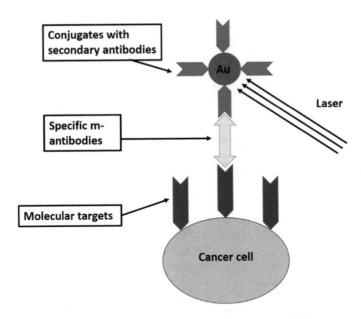

Fig. 4.8 Schematic representation of plasmonic photothermal therapy (PPTT)

Photothermal therapy is an invasive method in which photon energy is converted to heat to kill the tumour cells. Gold nanospheres, gold nano shells, gold nanorods based PTT are emerged as new technologies due their improved biocompatibility and photostability. Au NPs can absorb light 1000-fold greater than dye molecules ad the absorbed light converted into heat through nonradiative properties. In skin cancer, the properties of spherical Au NPs are employed due to SPR absorption in visible region by the help of visible pulsed lasers. The basic principles of PPTT are explained in Fig. 4.8. The molecular cancer targets attached with the cancer cells labelled with monoclonal specific antibodies(m-antibodies). These primary antibodies are sandwiched between the secondary antibodies conjugated Au nanoparticle and molecular targets. In the last step, the pulsed laser irradiates and this result drastic heating of Au NPs, thus it kills the cancer cells. The nano pulsed laser irradiation is highly selective and specific to the tumours and controlled localised damage of cells possible depending on the pulse duration and the particle size.

In 2003, Hirsch et al. described the NIR PPT method using PEGylated gold nano shells in breast carcinoma cells in vivo and in vitro studies [126]. Similarly, El-Sayed et al. [113] demonstrated the PTT using gold nanorods conjugated by anti-EGFR antibodies. They are specifically bind to the ENT cancer cells. In the treatment laser wavelength is overlapped with the SPR bands of Au nanorods and the labelled cancer cells are photothermally damaged easily under the laser exposure for 4 min without damaging the normal cells. As compared to the gold nano shells, gold nanorods gives effective treatment [35, 127, 128].

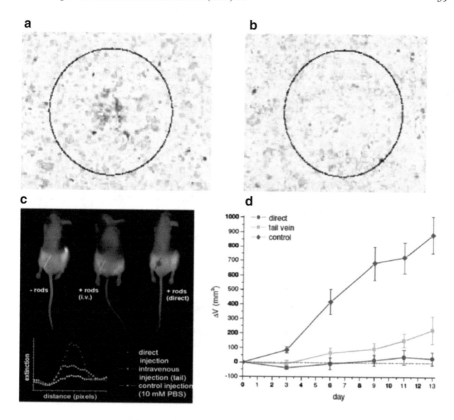

Fig. 4.9 a and **b** Selective in vitro photothermal cancer therapy using gold nanorods. Under NIR laser treatment (800 nm, 10 W/cm², 4 min), while the HSC-3 cancer cells undergo irreversible photo destruction indicated by tryphan blue staining (**a**), the HaCat normal cells are not affected (**b**). **c** In vivo NIR tumour imaging and **d** In vivo photothermal tumour therapy using gold nanorods [35]. Reproduced with permission from [45], [129]

Recently, Yue He et al. demonstrated the PTT in brain tumour tissues using the mixture of Au nano shells and Ag nanoparticles to reducing the side effects. The invitro photothermal cancer therapy using gold nanoparticles are shown in Fig. 4.9. The combination of plasmonic nanoparticles and Near Infra-Red laser can be utilised for the treatment of cancer.

4.3   Conclusion

This chapter has provided an overview of several key points of SPR and SERS. SPR theory, properties, factors affecting SPR/SERS and applications are described in the chapter. SERS also have widespread application as an analytical tool for the detection of diseases. The SERS techniques have been developed with the new types of substrates and instrumentation techniques. The combination of simultaneous SPR and SERS spectroscopy will give interesting aspects in analytical studies. The mechanism of plasmonic biosensing and the applications are narrated in the next chapter.

References

1. U. Kreibig, M. Vollmer, *Opticals Properties of Metal Clusters* (Springer, Berlin [u.a.], 1994)
2. G. Mie, Articles on the optical characteristics of turbid tubes, especially colloidal metal solutions. Ann. Phys. **25**, 377–445 (1908)
3. I. Angelini, G. Artioli, P. Bellintani et al., Chemical analyses of Bronze Age glasses from Frattesina di Rovigo, Northern Italy. J. Archaeol. Sci. **31**, 1175–1184 (2004). https://doi.org/10.1016/j.jas.2004.02.015
4. D.J. Barber, I.C. Freestone, An investigation of the origin of the colour of the Lycurgus cup by analytical transmission electron microscopy. Archaeometry **32**, 33–45 (1990). https://doi.org/10.1111/j.1475-4754.1990.tb01079.x
5. I. Freestone, N. Meeks, M. Sax, C. Higgitt, The Lycurgus cup—A Roman nanotechnology. Gold Bull. **40**, 270–277 (2007). https://doi.org/10.1007/BF03215599
6. P. Colomban, G. March, L. Mazerolles et al., Raman identification of materials used for jewellery and mosaics in Ifriqiya. J. Raman Spectrosc. **34**, 205–213 (2003). https://doi.org/10.1002/jrs.977
7. P. Colomban, The use of metal nanoparticles to produce yellow, red and iridescent colour, from bronze age to present times in lustre pottery and glass: solid state chemistry, spectroscopy and nanostructure. J. Nano Res. **8**, 109–132 (2009). https://doi.org/10.4028/scientific.net/JNanoR.8.109
8. S. Pérez-Villar, J. Rubio, J.L. Oteo, Study of color and structural changes in silver painted medieval glasses. J. Non Cryst. Solids **354**, 1833–1844 (2008). https://doi.org/10.1016/j.jnoncrysol.2007.10.008
9. F. Rubio, S. Pérez-Villar, M.Á. Garrido-Maneiro et al., Application of gradient and confocal Raman spectroscopy to analyze silver nanoparticle diffusion in medieval glasses. J. Nano Res. **8**, 89–97 (2009). https://doi.org/10.4028/scientific.net/JNanoR.8.89
10. S. Padovani, C. Sada, P. Mazzoldi et al., Copper in glazes of Renaissance luster pottery: nanoparticles, ions, and local environment. J. Appl. Phys. **93**, 10058–10063 (2003). https://doi.org/10.1063/1.1571965
11. M. Vilarigues, P. Fernandes, L.C. Alves, R.C. da Silva, Stained glasses under the nuclear microprobe: a window into history. Nucl. Instrum. Methods Phys. Res. Sect. B Beam Interact with Mater. Atoms **267**, 2260–2264 (2009). https://doi.org/10.1016/j.nimb.2009.03.049
12. Bobin, O., Schvoerer, M., Ney, C., et al., The role of copper and silver in the colouration of metallic luster decorations (Tunisia, 9th century; Mesopotamia, 10th century; Sicily, 16th century): a first approach. Color Res. Appl. **28**, 352–359 (2003). https://doi.org/10.1002/col.10183
13. S.K. Ghosh, T. Pal, Interparticle coupling effect on the surface plasmon resonance of gold nanoparticles: from theory to applications. Chem. Rev. **107**, 4797–4862 (2007). https://doi.org/10.1021/cr0680282
14. M. Ganguly, A. Pal, Y. Negishi, T. Pal, Synthesis of highly fluorescent silver clusters on gold(I) surface. Langmuir **29**, 2033–2043 (2013). https://doi.org/10.1021/la304835p
15. M. Brack, The physics of simple metal clusters: self-consistent jellium model and semiclassical approaches. Rev. Mod. Phys. **65**, 677–732 (1993). https://doi.org/10.1103/RevModPhys.65.677

16. C. Caucheteur, T. Guo, J. Albert, Review of plasmonic fiber optic biochemical sensors: improving the limit of detection. Anal. Bioanal. Chem. **407**, 3883–3897 (2015). https://doi.org/10.1007/s00216-014-8411-6
17. J. Jana, M. Ganguly, T. Pal, Enlightening surface plasmon resonance effect of metal nanoparticles for practical spectroscopic application. RSC Adv. **6**, 86174–86211 (2016). https://doi.org/10.1039/c6ra14173k
18. M.V. Bashevoy, F. Jonsson, A.V. Krasavin et al., Generation of traveling surface plasmon waves by free-electron impact. Nano Lett. **6**, 1113–1115 (2006). https://doi.org/10.1021/nl0 60941v
19. T. Nagao, T. Hildebrandt, M. Henzler, S. Hasegawa, Dispersion and damping of a two-dimensional plasmon in a metallic surface-state band. Phys. Rev. Lett. **86**, 5747–5750 (2001). https://doi.org/10.1103/PhysRevLett.86.5747
20. P. Johns, K. Yu, M.S. Devadas, G.V. Hartland, Role of resonances in the transmission of surface plasmon polaritons between nanostructures. ACS Nano **10**, 3375–3381 (2016). https://doi.org/10.1021/acsnano.5b07185
21. A. Otto, Excitation of nonradiative surface plasma waves in silver by the method of frustrated total reflection. Zeitschrift für Phys. A Hadron Nucl. **216**, 398–410 (1968). https://doi.org/10.1007/BF01391532
22. S. Zeng, D. Baillargeat, H.-P. Ho, K.-T. Yong, Nanomaterials enhanced surface plasmon resonance for biological and chemical sensing applications. Chem. Soc. Rev. **43**, 3426–3452 (2014). https://doi.org/10.1039/C3CS60479A
23. S. Wei, Q. Wang, J. Zhu et al., Multifunctional composite core–shell nanoparticles. Nanoscale **3**, 4474–4502 (2011). https://doi.org/10.1039/C1NR11000D
24. V. Bonacic-Koutecky, P. Fantucci, J. Koutecky, Quantum chemistry of small clusters of elements of groups Ia, Ib, and IIa: fundamental concepts, predictions, and interpretation of experiments. Chem. Rev. **91**, 1035–1108 (1991). https://doi.org/10.1021/cr00005a016
25. K.-P. Charlé, F. Frank, W. Schulze, The optical properties of silver microcrystallites in dependence on size and the influence of the matrix environment. Berichte der Bunsengesellschaft für Phys. Chemie **88**, 350–354 (1984). https://doi.org/10.1002/bbpc.19840880407
26. A. Adams, R.W. Rendell, W.P. West et al., Luminescence and nonradiative energy transfer to surfaces. Phys. Rev. B **21**, 5565–5571 (1980). https://doi.org/10.1103/PhysRevB.21.5565
27. L. Novotny, B. Hecht, *Principles of Nano-Optics* (Cambridge University Press, 2012)
28. M.S.A. Plasmonics, *Fundamentals and Applications* (Springer, Berlin, 2007)
29. A. Adams, J. Moreland, P.K. Hansma, Angular resonances in the light emission from atoms near a grating. Surf. Sci. **111**, 351–357 (1981)
30. A. Adams, J. Moreland, P.K. Hansma, Z. Schlesinger, Light emission from surface-plasmon and waveguide modes excited by N atoms near a silver grating. Phys. Rev. B **25**, 3457 (1982)
31. L.M. Liz-Marzán, *Nanometals: Formation and Color* (2004)
32. P.K. Jain, X. Huang, I.H. El-Sayed, M.A. El-Sayed, Review of some interesting surface plasmon resonance-enhanced properties of noble metal nanoparticles and their applications to biosystems. Plasmonics **2**, 107–118 (2007). https://doi.org/10.1007/s11468-007-9031-1
33. S. Link, M.A. El-Sayed, Optical properties and ultrafast dynamics of metallic nanocrystals. Annu. Rev. Phys. Chem. **54**, 331–366 (2003)
34. C. Sönnichsen, T. Franzl, T. Wilk et al., Drastic reduction of plasmon damping in gold nanorods. Phys. Rev. Lett. **88**, 77402 (2002)
35. X. Huang, M.A. El-Sayed, Gold nanoparticles: optical properties and implementations in cancer diagnosis and photothermal therapy. J. Adv. Res. **1**, 13–28 (2010). https://doi.org/10.1016/j.jare.2010.02.002
36. S. Link, M.A. El-Sayed, Spectral properties and relaxation dynamics of surface plasmon electronic oscillations in gold and silver nanodots and nanorods. J. Phys. Chem. B **103**, 8410–8426 (1999)
37. J. Cao, T. Sun, K.T.V. Grattan, Gold nanorod-based localized surface plasmon resonance biosensors: a review. Sens. Actuators, B Chem. **195**, 332–351 (2014). https://doi.org/10.1016/j.snb.2014.01.056

38. X. Huang, S. Neretina, M.A. El-Sayed, Gold nanorods: from synthesis and properties to biological and biomedical applications. Adv. Mater. **21**, 4880–4910 (2009)
39. V. Sharma, K. Park, M. Srinivasarao, Colloidal dispersion of gold nanorods: historical background, optical properties, seed-mediated synthesis, shape separation and self-assembly. Mater. Sci. Eng. R Rep. **65**, 1–38 (2009)
40. Lee KS, El-Sayed MA, Dependence of the enhanced optical scattering efficiency relative to that of absorption for gold metal nanorods on aspect ratio, size, end-cap shape, and medium refractive index. J. Phys. Chem. B. **109**, 20331–20338 (2005). https://doi.org/10.1021/jp0 54385p
41. L. Vigderman, B.P. Khanal, E.R. Zubarev, Functional gold nanorods: synthesis, self-assembly, and sensing applications. Adv. Mater. **24**, 4811–4841 (2012)
42. C.J. Murphy, A.M. Gole, S.E. Hunyadi et al., Chemical sensing and imaging with metallic nanorods. Chem. Commun. **8**, 544–557 (2002). https://doi.org/10.1039/b711069c
43. X. Huang, I.H. El-Sayed, W. Qian, M.A. El-Sayed, Cancer cell imaging and photothermal therapy in the near-infrared region by using gold nanorods. J. Am. Chem. Soc. **128**, 2115–2120 (2006). https://doi.org/10.1021/ja057254a
44. J.N. Anker, W.P. Hall, O. Lyandres, et al., Biosensing with plasmonic nanosensors. Nanoscience and Technology: A Collection of Reviews from Nature Journals. World Scientific (2009), pp. 308–319
45. E. Hutter, J.H. Fendler, Exploitation of localized surface plasmon resonance. Adv. Mater. **16**, 1685–1706 (2004). https://doi.org/10.1002/adma.200400271
46. R. Pilot, R. Signorini, L. Fabris, Surface-enhanced Raman spectroscopy: principles, substrates, and applications. *Metal Nanoparticles and Clusters* (Springer, 2018), pp. 89–164
47. B. Gilad, Entrepreneurship: the issue of creativity in the market place. J. Creat. Behav.
48. M. Fleischmann, P.J. Hendra, A.J. McQuillan, Raman spectra of pyridine adsorbed at a silver electrode. Chem. Phys. Lett. **26**, 163–166 (1974). https://doi.org/10.1016/0009-2614(74)853 88-1
49. M. Moskovits, Persistent misconceptions regarding SERS. Phys. Chem. Chem. Phys. **15**, 5301–5311 (2013). https://doi.org/10.1039/c2cp44030j
50. D.L. Jeanmaire, R.P. Van Duyne, Surface Raman spectroelectrochemistry. Part I. Heterocyclic, aromatic, and aliphatic amines adsorbed on the anodized silver electrode. J. Electroanal. Chem. **84**, 1–20 (1977). https://doi.org/10.1016/S0022-0728(77)80224-6
51. M.G. Albrecht, J.A. Creighton. J. Am. Chem. Soc. 99 (1977)
52. P. Hendra, THE DISCOVERY of SERS: An idiosyncratic account from a vibrational spectroscopist. Analyst **141**, 4996–4999 (2016). https://doi.org/10.1039/c6an90055k
53. M.G. Albrecht, J.A. Creighton, Anomalously intense Raman spectra of pyridine at a silver electrode. J. Am. Chem. Soc. **99**, 5215–5217 (1977)
54. S. Nie, S.R. Emory, Probing single molecules and single nanoparticles by surface-enhanced Raman scattering. Science (80-) **275**, 1102–1106 (1997). https://doi.org/10.1126/science.275. 5303.1102
55. E.J. Blackie, E.C. Le Ru, P.G. Etchegoin, Single-molecule surface-enhanced Raman spectroscopy of nonresonant molecules. J. Am. Chem. Soc. **131**, 14466–14472 (2009)
56. K.C. Bantz, A.F. Meyer, N.J. Wittenberg et al., Recent progress in SERS biosensing. Phys. Chem. Chem. Phys. **13**, 11551–11567 (2011). https://doi.org/10.1039/c0cp01841d
57. I. Ros, T. Placido, V. Amendola et al., SERS properties of gold nanorods at resonance with molecular, transverse, and longitudinal plasmon excitations. Plasmonics **9**, 581–593 (2014). https://doi.org/10.1007/s11468-014-9669-4
58. M. Kahraman, E.R. Mullen, A. Korkmaz, S. Wachsmann-Hogiu, Fundamentals and applications of SERS-based bioanalytical sensing. Nanophotonics **6**, 831–852 (2017). https://doi. org/10.1515/nanoph-2016-0174
59. Y.H. Wu, S. Sasaki, H. Shimizu, J. Raman Spectrosc. **26**, 963 (2003)
60. G.F. Walsh, L. Dal Negro, Enhanced second harmonic generation by photonic-plasmonic fano-type coupling in nanoplasmonic arrays. Nano Lett. **13**, 3111–3117 (2013). https://doi. org/10.1021/nl401037n

61. B. Sharma, R.R. Frontiera, A.-I. Henry et al., SERS: materials, applications, and the future. Mater Today **15**, 16–25 (2012)
62. C.L. Haynes, A.D. McFarland, R.P. Van Duyne, Surface-enhanced Raman spectroscopy (2005)
63. J. Grand, M.L. De La Chapelle, J.L. Bijeon et al., Role of localized surface plasmons in surface-enhanced Raman scattering of shape-controlled metallic particles in regular arrays. Phys. Rev. B—Condens. Matter Mater. Phys. **72**, 33407 (2005). https://doi.org/10.1103/PhysRevB.72.033407
64. C.L. Haynes, R.P. Van Duyne, Plasmon-sampled surface-enhanced Raman excitation spectroscopy. J. Phys. Chem. B **107**, 7426–7433 (2003). https://doi.org/10.1021/jp027749b
65. N. Félidj, J. Aubard, G. Lévi et al., Controlling the optical response of regular arrays of gold particles for surface-enhanced Raman scattering. Phys. Rev. B—Condens. Matter Mater. Phys. **65**, 0754191–0754199 (2002). https://doi.org/10.1103/PhysRevB.65.075419
66. A.D. McFarland, M.A. Young, J.A. Dieringer, R.P. Van Duyne, Wavelength-scanned surface-enhanced Raman excitation spectroscopy. J. Phys. Chem. B **109**, 11279–11285 (2005). https://doi.org/10.1021/jp050508u
67. A.J. Haes, C.L. Haynes, A.D. McFarland et al., Plasmonic materials for surface-enhanced sensing and spectroscopy. MRS Bull. **30**, 368–375 (2005). https://doi.org/10.1557/mrs2005.100
68. S.K. Gray, T. Kupka, Propagation of light in metallic nanowire arrays: finite-difference time-domain studies of silver cylinders. Phys. Rev. B—Condens. Matter Mater. Phys. **68**, 454151–4541511 (2003). https://doi.org/10.1103/PhysRevB.68.045415
69. J.A. Dieringer, A.D. McFarland, N.C. Shah et al., Introductory lecture surface enhanced Raman spectroscopy: new materials, concepts, characterization tools, and applications. Faraday Discuss. **132**, 9–26 (2006). https://doi.org/10.1039/B513431P
70. A.V. Whitney, J.W. Elam, S. Zou et al., Localized surface plasmon resonance nanosensor: a high-resolution distance-dependence study using atomic layer deposition. J. Phys. Chem. B **109**, 20522–20528 (2005). https://doi.org/10.1021/jp0540656
71. X. Lu, M. Rycenga, S.E. Skrabalak et al., Chemical synthesis of novel plasmonic nanoparticles. Annu. Rev. Phys. Chem. **60**, 167–192 (2009). https://doi.org/10.1146/annurev.physchem.040808.090434
72. T. Ding, D.O. Sigle, L.O. Herrmann et al., Nanoimprint lithography of Al nanovoids for deep-UV SERS. ACS Appl. Mater. Interfaces **6**, 17358–17363 (2014). https://doi.org/10.1021/am505511v
73. M. Rycenga, C.M. Cobley, J. Zeng et al., Controlling the synthesis and assembly of silver nanostructures for plasmonic applications. Chem. Rev. **111**, 3669–3712 (2011). https://doi.org/10.1021/cr100275d
74. Njoki PN, Lim IIS, Mott D, et al., Size correlation of optical and spectroscopic properties for gold nanoparticles. J. Phys. Chem. C. **111**, 14664–14669 (2007). https://doi.org/10.1021/jp074902z
75. Ou S, Xie Q, Ma D, et al., A precursor decomposition route to polycrystalline CuS nanorods. Mater. Chem. Phys. **94**, 460–466 (2005). https://doi.org/https://doi.org/10.1016/j.matchemphys.2005.04.057
76. Kahraman M, Daggumati P, Kurtulus O, et al., Fabrication and characterization of flexible and tunable plasmonic nanostructures. Sci. Rep. **3**, 1–9 (2013). https://doi.org/10.1038/srep03396
77. Orendorff CJ, Gole A, Sau TK, Murphy CJ, Surface-enhanced Raman spectroscopy of self-assembled monolayers: Sandwich architecture and nanoparticle shape dependence. Anal. Chem. **77**, 3261–3266 (2005). https://doi.org/10.1021/ac048176x
78. Wiley BJ, Chen Y, Mclellan JM, et al., (2007) 1368.NanoLett.2007,7,1032.pdf
79. Wu HY, Choi CJ, Cunningham BT, Plasmonic nanogap-enhanced raman scattering using a resonant nanodome array. Small **8**, 2878–2885 (2012). https://doi.org/10.1002/smll.201200712

80. Ye J, Wen F, Sobhani H, et al., Plasmonic nanoclusters: Near field properties of the Fano resonance interrogated with SERS. Nano Lett **12**, 1660–1667 (2012). https://doi.org/10.1021/nl3000453

81. Singh JP, Chu H, Abell J, et al., Flexible and mechanical strain resistant large area SERS active substrates. Nanoscale **4**, 3410–3414 (2012). https://doi.org/10.1039/c2nr00020b

82. Chung AJ, Huh YS, Erickson D, Large area flexible SERS active substrates using engineered nanostructures. Nanoscale **3**, 2903–2908 (2011). https://doi.org/10.1039/c1nr10265f

83. V. Amendola, R. Pilot, M. Frasconi, et al., Surface plasmon resonance in gold nanoparticles: a review. J. Phys. Condens. Matter **29**, 203002 (2017). https://doi.org/10.1088/1361-648X/aa60f3

84. L. Rivas, S. Sanchez-Cortes, J.V. García-Ramos, G. Morcillo, Mixed silver/gold colloids: a study of their formation, morphology, and surface-enhanced Raman activity. Langmuir **16**, 9722–9728 (2000). https://doi.org/10.1021/la000557s

85. Y. Yang, J. Shi, G. Kawamura, M. Nogami, Preparation of Au–Ag, Ag–Au core–shell bimetallic nanoparticles for surface-enhanced Raman scattering. Scr. Mater. **58**, 862–865 (2008). https://doi.org/10.1016/j.scriptamat.2008.01.017

86. Y. Huang, Y. Yang, Z. Chen et al., Fabricating Au–Ag core-shell composite films for surface-enhanced Raman scattering. J. Mater. Sci. **43**, 5390–5393 (2008). https://doi.org/10.1007/s10853-008-2793-9

87. J.B. Jackson, N.J. Halas, Surface-enhanced Raman scattering on tunable plasmonic nanoparticle substrates. Proc. Natl. Acad. Sci. **101**, 17930 LP–17935 (2004). https://doi.org/10.1073/pnas.0408319102

88. T.L. Moore, L. Rodriguez-Lorenzo, V. Hirsch et al., Nanoparticle colloidal stability in cell culture media and impact on cellular interactions. Chem. Soc. Rev. **44**, 6287–6305 (2015). https://doi.org/10.1039/C4CS00487F

89. A.I. Pérez-Jiménez, D. Lyu, Z. Lu et al., Surface-enhanced Raman spectroscopy: benefits, trade-offs and future developments. Chem. Sci. **11**, 4563–4577 (2020). https://doi.org/10.1039/D0SC00809E

90. M.R. Bailey, A.M. Pentecost, A. Selimovic et al., Sheath-flow microfluidic approach for combined surface enhanced Raman scattering and electrochemical detection. Anal. Chem. **87**, 4347–4355 (2015). https://doi.org/10.1021/acs.analchem.5b00075

91. B.L. Scott, K.T. Carron, Dynamic surface enhanced Raman spectroscopy (SERS): extracting SERS from normal Raman scattering. Anal. Chem. **84**, 8448–8451 (2012). https://doi.org/10.1021/ac301914a

92. R.A. Álvarez-Puebla, Effects of the excitation wavelength on the SERS spectrum. J. Phys. Chem. Lett. **3**, 857–866 (2012). https://doi.org/10.1021/jz201625j

93. R.A. Sperling, W.J. Parak, Surface modification, functionalization and bioconjugation of colloidal inorganic nanoparticles. Philos. Trans. R Soc. A Math. Phys. Eng. Sci. **368**, 1333–1383 (2010). https://doi.org/10.1098/rsta.2009.0273

94. W. Xi, B.K. Shrestha, A.J. Haes, Promoting intra- and intermolecular interactions in surface-enhanced Raman scattering. Anal. Chem. **90**, 128–143 (2018). https://doi.org/10.1021/acs.analchem.7b04225

95. N.G. Tognalli, A. Fainstein, E.J. Calvo et al., Incident wavelength resolved resonant SERS on Au Sphere Segment Void (SSV) arrays. J. Phys. Chem. C **116**, 3414–3420 (2012). https://doi.org/10.1021/jp211049u

96. S.H. Lee, K.C. Bantz, N.C. Lindquist et al., Self-assembled plasmonic nanohole arrays. Langmuir **25**, 13685–13693 (2009). https://doi.org/10.1021/la9020614

97. Q. Yu, P. Guan, D. Qin et al., Inverted size-dependence of surface-enhanced Raman scattering on gold nanohole and nanodisk arrays. Nano Lett. **8**, 1923–1928 (2008). https://doi.org/10.1021/nl0806163

98. A.G. Brolo, E. Arctander, R. Gordon et al., Nanohole-enhanced Raman scattering. Nano Lett. **4**, 2015–2018 (2004). https://doi.org/10.1021/nl048818w

99. A.J. Baca, T.T. Truong, L.R. Cambrea, et al., Molded plasmonic crystals for detecting and spatially imaging surface bound species by surface-enhanced Raman scattering. Appl. Phys. Lett. **94**, 243109 (2009). https://doi.org/10.1063/1.3155198

100. N.A. Cinel, S. Bütün, G. Ertaş, E. Özbay, 'Fairy Chimney'-shaped tandem metamaterials as double resonance SERS substrates. Small **9**, 531–537 (2013). https://doi.org/10.1002/smll. 201201286

101. Q. Yu, S. Braswell, B. Christin, et al., Surface-enhanced Raman scattering on gold quasi-3D nanostructure and 2D nanohole arrays. Nanotechnology **21**, 355301 (2010). https://doi.org/10.1088/0957-4484/21/35/355301

102. M. Ranjan, S. Facsko, Anisotropic surface enhanced Raman scattering in nanoparticle and nanowire arrays. Nanotechnology **23**, 485307 (2012). https://doi.org/10.1088/0957-4484/23/48/485307

103. M.K. Hossain, K. Shibamoto, K. Ishioka, et al., 2D nanostructure of gold nanoparticles: an approach to SERS-active substrate. J. Lumin. **122–123**, 792–795 (2007). https://doi.org/10.1016/j.jlumin.2006.01.290

104. C. Wang, F. Madiyar, C. Yu, J. Li, Detection of extremely low concentration waterborne pathogen using a multiplexing self-referencing SERS microfluidic biosensor. J. Biol. Eng. **11**, 1–11 (2017)

105. C. Song, B. Yang, Y. Zhu et al., Ultrasensitive sliver nanorods array SERS sensor for mercury ions. Biosens. Bioelectron. **87**, 59–65 (2017)

106. M.J. Oliveira, P. Quaresma, M.P. de Almeida et al., Office paper decorated with silver nanostars—an alternative cost effective platform for trace analyte detection by SERS. Sci. Rep. **7**, 1–14 (2017)

107. X. Li, S. Feng, Y. Hu et al., Rapid detection of melamine in milk using immunological separation and surface enhanced Raman spectroscopy. J. Food Sci. **80**, C1196–C1201 (2015)

108. H. Qiu, M. Wang, S. Jiang et al., Reliable molecular trace-detection based on flexible SERS substrate of graphene/Ag-nanoflowers/PMMA. Sensors Actuators B Chem **249**, 439–450 (2017)

109. N. Chen, P. Ding, Y. Shi et al., Portable and reliable surface-enhanced Raman scattering silicon chip for signal-on detection of trace trinitrotoluene explosive in real systems. Anal. Chem. **89**, 5072–5078 (2017)

110. O.S. Kulakovich, E.V. Shabunya-Klyachkovskaya, A.S. Matsukovich et al., Nanoplasmonic Raman detection of bromate in water. Opt. Express **24**, A174–A179 (2016)

111. M.D. Malinsky, K.L. Kelly, G.C. Schatz, R.P. Van Duyne, Chain length dependence and sensing capabilities of the localized surface plasmon resonance of silver nanoparticles chemically modified with alkanethiol self-assembled monolayers. J. Am. Chem. Soc. **123**, 1471–1482 (2001). https://doi.org/10.1021/ja003312a

112. N.G. Khlebtsov, V.A. Bogatyrev, B.N. Khlebtsov et al., A multilayer model for gold nanoparticle bioconjugates: application to study of gelatin and human IgG adsorption using extinction and light scattering spectra and the dynamic light scattering method. Colloid J. **65**, 622–635 (2003). https://doi.org/10.1023/A:1026140310601

113. I.H. El-Sayed, X. Huang, M.A. El-Sayed, Surface plasmon resonance scattering and absorption of anti-EGFR antibody conjugated gold nanoparticles in cancer diagnostics: applications in oral cancer. Nano Lett. **5**, 829–834 (2005). https://doi.org/10.1021/nl050074e

114. D.A. Schultz, Plasmon resonant particles for biological detection. Curr. Opin. Biotechnol. **14**, 13–22 (2003). https://doi.org/10.1016/S0958-1669(02)00015-0

115. C. Loo, L. Hirsch, M.-H. Lee et al., Gold nanoshell bioconjugates for molecular imaging in living cells. Opt. Lett. **30**, 1012–1014 (2005). https://doi.org/10.1364/OL.30.001012

116. R. Hu, K.-T. Yong, I. Roy et al., Metallic nanostructures as localized plasmon resonance enhanced scattering probes for multiplex dark-field targeted imaging of cancer cells. J. Phys. Chem. C **113**, 2676–2684 (2009). https://doi.org/10.1021/jp8076672

117. M. Eghtedari, A.V. Liopo, J.A. Copland et al., Engineering of hetero-functional gold nanorods for the in vivo molecular targeting of breast cancer cells. Nano Lett. **9**, 287–291 (2009). https://doi.org/10.1021/nl802915q

118. L.A. Austin, B. Kang, C.-W. Yen, M.A. El-Sayed, Plasmonic imaging of human oral cancer cell communities during programmed cell death by nuclear-targeting silver nanoparticles. J. Am. Chem. Soc. **133**, 17594–17597 (2011). https://doi.org/10.1021/ja207807t

119. S. Lal, S.E. Clare, N.J. Halas, Nanoshell-enabled photothermal cancer therapy: impending clinical impact. Acc. Chem. Res. **41**, 1842–1851 (2008). https://doi.org/10.1021/ar800150g

120. G. Han, P. Ghosh, V.M. Rotello, Functionalized gold nanoparticles for drug delivery. Nanomedicine **2**, 113–123 (2007). https://doi.org/10.2217/17435889.2.1.113

121. R. Bhattacharya, P. Mukherjee, Biological properties of "naked" metal nanoparticles. Adv. Drug Deliv. Rev. **60**, 1289–1306 (2008). https://doi.org/10.1016/j.addr.2008.03.013

122. P. Ghosh, G. Han, M. De, et al., Gold nanoparticles in delivery applications. Adv. Drug Deliv. Rev. **60**, 1307–1315 (2008). https://doi.org/10.1016/j.addr.2008.03.016

123. G.F. Paciotti, D.G.I. Kingston, L. Tamarkin, Colloidal gold nanoparticles: a novel nanoparticle platform for developing multifunctional tumor-targeted drug delivery vectors. Drug Dev. Res. **67**, 47–54 (2006). https://doi.org/10.1002/ddr.20066

124. Y. Cheng, A.C. Samia, J.D. Meyers et al., Highly efficient drug delivery with gold nanoparticle vectors for in vivo photodynamic therapy of cancer. J. Am. Chem. Soc. **130**, 10643–10647 (2008). https://doi.org/10.1021/ja801631c

125. B. Khlebtsov, V. Zharov, A. Melnikov et al., Optical amplification of photothermal therapy with gold nanoparticles and nanoclusters. Nanotechnology **17**, 5167–5179 (2006). https://doi.org/10.1088/0957-4484/17/20/022

126. L.R. Hirsch, R.J. Stafford, J.A. Bankson, et al., Nanoshell-mediated near-infrared thermal therapy of tumors under magnetic resonance guidance. Proc. Natl. Acad. Sci. **100**, 13549 LP–13554 (2003). https://doi.org/10.1073/pnas.2232479100

127. C.M. Pitsillides, E.K. Joe, X. Wei, et al., Selective cell targeting with light-absorbing microparticles and nanoparticles. Biophys. J. **84**, 4023–4032 (2003). https://doi.org/10.1016/S0006-3495(03)75128-5

128. H.P. Berlien, G.J. Müller, *Applied Laser Medicine* (Springer, 2003)

129. Dickerson EB, Dreaden EC, Huang X, et al., Gold nanorod assisted near-infrared plasmonic photothermal therapy (PPTT) of squamous cell carcinoma in mice. Cancer Lett. **269**, 57–66 (2008). https://doi.org/10.1016/j.canlet.2008.04.026

Chapter 5
Advances in Plasmonic Biosensors and Their Futuristic Applications

Neeli Chandran, Manikanta Bayal, Rajendra Pilankatta, and Swapna S. Nair

5.1 Introduction

The focus of this chapter is on different types of sensors based on plasmonic nanoparticles for detecting various analytes. These include biological and chemical entities including metal ions, organic molecules, proteins, biomarkers, nucleic acids (DNA and RNA) etc. Biosensors exhibits great potential in medical diagnosis, food safety and environmental monitoring [1–3]. When the biosensors detect the biomarkers of life-threatening diseases and harmful chemical components, it should provide high level of accuracy of detection and specificity. Early detection of diseases is possible by detecting the presence or variations of several biological components in the human body. These are achievable through the development of specific biosensors. Several reviews are available on the development of such type of chemi/biosensors. Here, we briefly discuss about the structure, sensing mechanisms, detection methods and applications of plasmonic biosensors.

5.2 Biosensors

Generally, biosensors employ the help of a suitable bioreceptor molecule as a recognition element. These are the molecules those particularly bind to the analyte. A transducer is another important element in the biosensor, which is often an interface,

N. Chandran · M. Bayal · S. S. Nair (✉)
Department of Physics, Central University of Kerala, Kasaragod 671320, India

R. Pilankatta
Department of Biochemistry and Molecular Biology, Central University of Kerala, Kasaragod 671320, India

© The Author(s), under exclusive license to Springer Nature Singapore Pte Ltd. 2021
S. S. Nair and R. Philip, *Nanomaterials for Luminescent Devices, Sensors, and Bio-imaging Applications*, Progress in Optical Science and Photonics 16,
https://doi.org/10.1007/978-981-16-5367-4_5

like a nanoparticle surface or electrodes, used to convert the analyte binding mechanisms into readable signal. The unique, optical, electrical and chemical properties of the materials are used for the detection of analytes. Nanomaterials based sensors has great potential in biosensing application, and hence they are intensively being explored in the biomedical field. The nano structured materials provide intelligent reaction mechanisms with the lowest limit of detection bettered by several order of magnitudes [4]. General advantage of nanomaterials is the highly specific surfaces to immobilize vast variety of biological receptors. Analytical biosensors can provide accurate concentration of the analyte with the linearity of the concentration of the analyte and the detected optical/electrical signal. Basic characteristics of biosensors are linearity, selectivity, sensitivity, wide range of response and the lowest limit of detection (LOD). etc. Based on the detected signals, biosensors are classified into optical biosensors, electrical biosensors piezoelectric biosensors. etc. [5]. Due to the high sensitivity, specificity and low volume of samples needed, optical biosensors are widely used for analytical biosensing [4, 6].

5.2.1 Mechanism of Sensing

As discussed above, biosensor is an analytical device to measure the concentration of analytes through chemical or biological reactions by generating signals proportional to them. Biosensors are employed for the disease detection, detection of toxic components and heavy metal ions, DNA detection virus detection etc. from the human body fluids. The mechanism of a biosensor is shown in the Fig. 5.1. It contains elements including bioreceptor, transducer component, electronics component and distal read out or display unit. A substance that needs detection is called **analyte**. A molecule that is specifically attached to the analyte is called **bioreceptor**. It may be organic molecules, peptides, proteins, polymers etc. These bioreceptors are attached to the analytes through covalent bonding, Vander Waals interactions, or through hydrogen bonding. The biorecognition process is the interaction between the bioreceptor and analyte which generate signals in the form of heat, light, pH change, mass change etc.

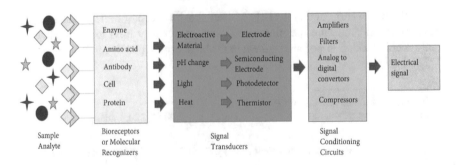

Fig. 5.1 Schematic representation of a biosensor

Most important component in the biosensor is **transducer**, which converts one form of energy into other. Transducer convert the biorecognition event into measurable form. Photodiode, and are the examples of transducers, which produces electrical or optical signals. These signals are proportional to the interaction between bioreceptors and analytes. A biosensor device completes by itself when an electronic component and a display unit is added with the sensor system. Electronic component converts the analogue signal to digital form.

5.2.2 Nanomaterials for Biosensing

A large variety of materials are used commonly for biosensing applications. Nanomaterials have enhanced performance in chemi/biosensors because they exhibit unique optical and electrical properties compared to their bulk counterparts. Semiconductor nanoparticles show bright fluorescence while metallic nanoparticles exhibit tuneable surface plasmon resonance (SPR) properties. Colloidal nanoparticles-based biosensors are a category of biological probe for detecting diseases and intra cellular imaging applications. Fluorescent semiconductor quantum dots and Au NPs are commonly used for biosensing applications because they give high quantum yield and excellent photostability [7, 8]. Their sharp absorption spectra are tuneable with particle size and shape. Here, we described about the optical biosensors, especially about the plasmonic biosensors. According to the detection methods, it can be classified into SPR based biosensors and SERS biosensors. In addition to this, absorption, reflection and transmission-based detection is also possible. Fluorescence based detection methods also have wide acceptance in biosensing.

5.3 Plasmonic Biosensors

The development and use of plasmonic biosensors for point of care diagnostic tools applications consists of different steps includes the fabrication of substrates, preparation of samples and detection of signals. Plasmonic materials include metal nanoparticles, graphene and other carbon-based quantum dots with different dimensions. Working principles are different for plasmonic biosensors according to their signal detection methods which is shown in the Fig. 5.2.

5.3.1 Surface Plasmon Resonance (SPR)

SPR based detection of biological entities are widely sought after as a label free detection technique to quantify the molecular interaction between the metallic nanoparticles using surface plasmons. When light interacts with the metal nanoparticles,

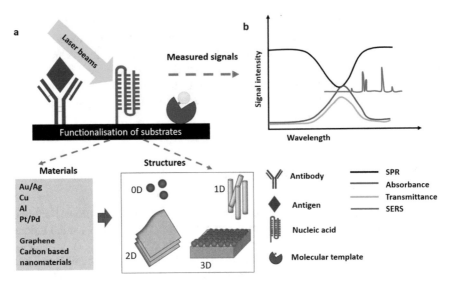

Fig. 5.2 Schematic representation of plasmonic biosensors **a** fabrication and working of biosensors with different plasmonic materials and **b** different types of detection methods. Image concept courtesy [6]

surface plasmon polaritons (SPP) are generated as described in the previous chapter. Noble metallic nanoparticles like gold and silver possess excellent surface plasmon resonance properties in the nano range as compared to its bulk form. When electromagnetic waves interact with the conduction elections on the surface of the nanoparticles, resonance occurs when frequencies match. A strong absorption band is often observed in the case of plasmonic particles. These peaks are tuneable with respect to the size, shape, dielectric medium and other changes in the environment. In nanoparticles, highly confined electromagnetic fields are induced at their surface by localised surface plasmon resonance, which is very much sensitive to the small changes in the dielectric environment. Due to this reason, LSPR sensors are widely used for biomedical applications. When the light interacts with the metallic substrate at a particular angle, the light is absorbed by the conduction electrons and a resonance condition is reached [9]. The light was absorbed by the conduction electrons at this SPR angle or resonance angle. This causes a sharp reduction of reflectance at a particular wavelength. When the refractive index of the surrounding medium changes, the reflected light (that are not absorbed) will be detected. This is caused by the absorption of metallic substrates by the interaction between the bioreceptors with the target molecules [10–13].

The mechanism of SPR and LSPR are shown in the Fig. 5.3. SPR based biosensing method uses the total internal reflection of light at the metal substrate -dielectric interface and the surface plasmon polaritons generated at certain angles. For an SPR based system, it includes the prism coupler and the electromagnetic wave propagating along metal-prism interface which depends on the refractive index of the prism and

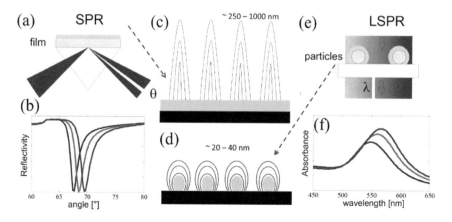

Fig. 5.3 The mechanism of SPR and LSPR. Reproduced with permission from [6] and [14]

the incident angle. LSPR based sensors are different from traditional SPR sensors based on metallic films. As discussed above, LSPR is generated at the surface of metallic nanoparticles with localised field when the decay length is much shorter. This confinement increases the electric field around the metallic nanoparticles. This type of LSPR is sensitive to the molecular binding interaction between the nanoparticles and small biological entities [14].

There are two types of surface plasmons; propagating and non-propagating. Propagating surface plasmon is known as surface plasmon polariton which is generated at the noble metallic thin film surface. Non propagating surface plasmons are localised plasmons which are generated at the surface of nanoparticles with 10–200 nm size range [15, 16]. There are few advantages for them which can be listed as follows.

(1)   The high sensitivity due to the refractive index changes.
(2)   Label free techniques are possible because of spectral shifts measurements.
(3)   Excellent reproducibility on SERS substrate from sensor to sensors and
(4)   Easy instrument setup and low-cost production are other advantages of LSPR sensors.

5.3.2 Sensing Strategies

The plasmonic biosensors directly utilize the surface plasmonic properties through different mechanisms. They utilize the shift in spectra of plasmon by transducing the sensing signal. Plasmonic biosensors utilize the mechanisms including localised surface plasmon resonance (LSPR), propagating surface plasmon (SPP), Surface enhanced Raman Spectroscopy (SERS) and Surface enhanced Fluorescence (SEF) etc. Details of each method are discussed in the following sections.

5.3.2.1   LSPR Based Biosensors

Two-dimensional plasmonic substrates perform with the propagating surface plasmon polaritons (SPP) mode or the combination of SPP/LSPR mode. This can happen by the changes in the refractive index of the medium. SPPs are the propagating plasmon waves on the surface of a metallic substrate which cannot be excited by the free space radiation. They can be modulated by the refractive index of the medium; thus, transducing the signal corresponding to particular biological interaction. SPP offers longer decay length than LSPR and hence it can be modulated by the changes occurred at the nanoparticle surface which is farther. a large variety of metallic/bi-metallic core shell nanomaterials are commonly used for making plasmonic biosensors. Among this, gold (Au) and Silver (Ag) nanoparticles and the substrates based on them are widely used for biosensing applications.

While coming to the basics of SPR, the electromagnetic light influences the conduction electrons in the metal. The resulting motion of the electrons are oscillating with 180° out of phase due to the negative charge of electrons. The dampening of oscillation occurs and this leads to ohmic loses [17]. The plasma frequency, which is the characteristic frequency of conduction electrons, depends on the effective mass and density of the conduction electrons.

If the incident light has a frequency greater than plasma frequency (in UV range in the case of metals), the electrons will not oscillate and the light is absorbed or transmitted in inter-band transitions. If the frequency is smaller than UV light, the electrons will start to oscillate at 180° out of phase and resulting strong reflection [17]. Metal particles achieves its characteristic colour by the combination of this plasma frequency and inter-band transitions. Colorimetric sensing with the help of metallic nanoparticles is possible due to this reason of plasmon resonances.

LSPR and SPP have some differences when they are discussed in the point of view of a sensor. The geometry of the nanoparticle significantly depends on the properties. As described in the previous chapter, LSPR positions are changed according to the size, shape and environment of the nanoparticles. In the case of metal nano surfaces, the electromagnetic decay length is of the order of 10–40 nm as compared to the SPR exhibiting continuous metal films for which it is in 100–400 nm range. This improves the surface sensitivity of plasmonic substrates for biosensing applications. Bulk sensitivity and temperature fluctuations are negligible in LSPR which enhances the nano surface sensitivity. SPR techniques require an external light coupling method; but in LSPR techniques, light is directly coupled with the sensor surface, which reduces the complexity of device fabrication. So, LSPR gives excellent potential for the development of portable clinical diagnostic devices.

Macroscale and nanoscale LSPR detection methods are in use depending on the number of observed particles. The chemical reduction of metallic salts in the presence of suitable surfactants is the most common method used in the case of synthesis of metallic nanoparticles. Adsorption of nanoparticles on to the LSPR substrate have some limitations including polydispersity and reproducibility. To overcome these limitations, lithographic techniques were used. Uniformity of size and shape of metal nanoparticles can be achieved by the lithographic techniques. Electron

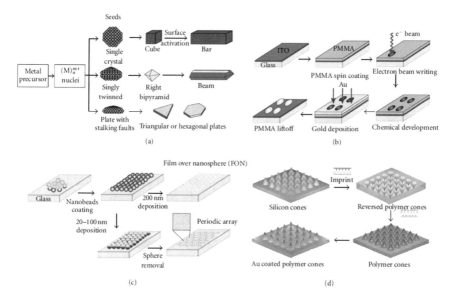

Fig. 5.4 Different types of substrate fabrication techniques. **a** Wet seed-mediated chemical reduction method. **b** Principle of electron beam technique. **c** Nanosphere lithography process. **d** Schematic of the nanofabrication process for 3D gold-coated nano cones by nanoimprint lithography. Reprinted with permission from [18]. Copyright 2011 Elsevier

beam lithography and ion beam lithography are also widely used for the fabrication of LSPR sensor substrates. Ion beam lithography produces 2D periodic arrays which are more sensitive to the detection of biomolecules. Macroscale detection method composes of three parts- a light source, large number of nanoparticles containing sample chip and a spectrometer. The light passes from the source will fall on the sample chip, which gives the information about the analyte adsorption [18]. Then it arrives at the spectrometer. Nanoscale detection method contains more exquisite and sophisticated systems. Figure 5.4 shows different substrate fabrication techniques.

Sensing of biomarkers of several life-threatening diseases are possible by the LSPR sensors. LSPR sensors can be used in biological assays to detect the biomolecules. Generally, the accurate quantification and estimation of biomarkers assume great importance in disease detection and theranostics. Immunoassay methods like fluorescence immunoassays produce reliable diagnostics, but the researchers focus on the accurate and early screening of molecular biomarkers which can offer early detection of diseases. This leads to the development of advanced and ultrasensitive detecting tools based on LSPR sensors. LSPR based nano biosensors offer powerful technologies with real time and ultrasensitive performance for the detection of biomolecules. To ensure the ability of sensitive detection of biomarkers, biotin-streptavidin interaction is commonly exploited in LSPR biosensors. Biotin is a small organic molecule which can be attached to the surface of nanoparticles. In contrast with this, streptavidin is a large molecule that can be easily detected by measuring the changes in the refractive index of the environment. The interaction between the biotin and streptavidin shows a spectral shift in the SPR spectra corresponding to the concentration of streptavidin in even as low as nanomolar concentration levels. LSPR sensors have been used as an immunoassay format, i.e., Antigen antibody interaction. This can be utilized in real time analysis. Specific interaction kinetics of the antigen and antibody possess reasonable binding rates, due to these reasons, LSPR sensors have been used to conduct real time analysis. LSPR biosensor for the biomarker detection of Alzheimer's diseases was developed by Haes et al., which can detect the picomolar concentration of amyloid-derived diffusible ligands (ADDLs) [19]. The building blocks of ADDL, was present in both reduced and oxidized forms, which is very important in the progress of Alzheimer's diseases.

Kim et al. recently developed another type of LSPR biosensor for the detection of hepatitis B [20]. It consists of Au NPs sandwich immunoassays by conjugating Au NPs with an anti-HBsAg antibody which is utilized to detect HBsAg antigen. It can be possible to sense the concentration of HBsAg even in pg/mL level through measuring the absorbance spectra of both conjugated and unconjugated Au NPs and their associated spectral shifts which is shown in the Fig. 5.5. Avian influence virus (AIV H5N1) also can be detected with the LSPR technology with the help of hollow Au spike like nanoparticles [21, 22].

LSPR sensors has also been used for the diagnosis of cancer [23]. It has been used for the early recognition of molecular biomarker of cancers and the monitoring of therapeutic efficacy. Single molecule nanoparticle based optical sensors, which are rather a controlled system, where the single antibody molecules are attached to the surface of individual nanoparticles. Such tiny sensors offer the detection of single biomolecules. The detection of nuclear tumour suppressor protein, p53 is possible by LSPR biosensor based on triangular shaped silver nanoparticles in neck and head squamous cell carcinoma patients with the detection limit of 59.45 pg/mL [24]. Similarly, LSPR nanoparticles like Au nanosphere [25] and Ag nanosphere [26] are effectively used for the detection of prostate specific antigen (PSA), tumour necrosis factor etc.

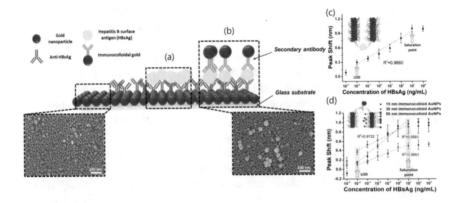

Fig. 5.5 LSPR biosensor working and analytical performance [21]. Reprinted with permission [20] [Copyright 2018 Elsevier]

5.3.2.2 Plasmon Enhanced Fluorescence Biosensors

Fluorescence based sensing methods are used in analytical chemistry for the detection of metal ions, toxic molecular components and biotechnology/ biochemistry for labelling of specific biomarkers. The interaction between the fluorophores and the free electrons presents in the plasmonic substrates or plasmonic nanoparticles depends on the fluorescence properties of the materials. Generally, the interaction between them can alter the fluorescence by enhancement and quenching depends upon the parameters like distance between the fluorophore and plasmonic nanoparticles [27], size and shape of the materials [28], the orientation of fluorophores [29] etc. This type of surface plasmon enhanced fluorescence are commonly used for biosensing applications.

Generally, photoluminescence occurs when electrons are excited by the incident photons for a certain time and then go through the radiative relaxation. Thus, the photons are reemitted when electrons fall into the ground state [30]. The photoluminescent properties are originated either from the intrinsic properties of the materials or from external analytes/fluorophore labels that are tagged to them. Intrinsic luminescence properties are very weak with shorter life time. Moreover, metal nanoparticles show very feeble luminescence properties. By adding fluorophore tags, we can induce fluorescence with high quantum yield [30]. Fluorescence based optical biosensors have been widely used for the detection of biomolecular analytes. But they have very shorter life time such as ns/ sub ns range [31]. Fluorescence signal of the targeted analytes can be enhanced by coupling it with surface plasmon resonance active substrates [32]. Rate of radiative and non-radiative decay depend on the fluorescence quantum yield. When a fluorophore is present in the vicinity of the surface of plasmonic materials, plasmons influence the radiative and nonradiative decay. The interaction between the fluorophore and propagating surface plasmons or LSPs can enhance the fluorescence intensity by increasing the quantum yield [33]. Otherwise, the excitation process can be influenced by plasmonically enhanced electromagnetic fields near the surface of nanoparticles. This may increase the intensity of light absorbed by the fluorophores. Plasmon enhances fluorescence properties that usually occurs at a specific distance of 5–30 nm. When it is too closer, the fluorescence is quenched, i.e., the emission peak of fluorophore is blue shifted with respect to the SPR band of metallic nanoparticle. The spectra are red shifted when the fluorescence enhanced [34, 35].

Semiconductor quantum dots draw much research attention due to their broad absorption spectra and narrow emission spectra. Among them, fluorescence quantum dots are used for biosensing applications. LSPR based sensors often exploits the biosensing applications based on quantum dots because, the presence of surface plasmon active metal nanoparticles result in the enhancement of fluorescence signals in quantum dots. The fluorescence is usually quenched according to the increasing concentration of analyte. Surface plasmon enhanced fluorescence-based biosensors are widely used for the detection of biological samples including viruses [21]. Takemura et al. introduced an LSPR immunosensor for the detection of H1N1 influence virus, with the help of CdSeTeS QDs which generated fluorescent signals [36]. These

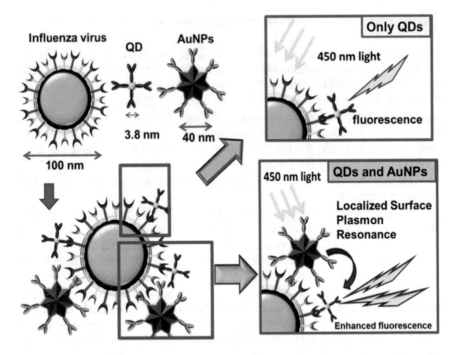

Fig. 5.6 The schematic representation of detection principle of H1N1influenza virus by surface plasmon induced fluorescence based nano plasmonic biosensor [36]. Reprinted with permission. Copyright 2008 Elsevier

fluorescence enhancements can be triggered by the LSPR of Au NPs which allowed specific detection of antigens on the surface of H1N1 virus which is shown in the Fig. 5.6. The QD's fluorescence intensity is quenched owing to the steric hindrance of LSPR of Au NPs because of the presence of influenza virus and the effect is linear against the virus concentration which makes it possible for accurate determination of virus concentration [36].

5.4 Application

As discussed in the previous sections, plasmonic biosensors are widely employed as point of care (POC) diagnostic devices in clinical uses. The selective and sensitive detection of biomarkers of several diseases can be possible through the plasmonic biosensors. biomarker detection of neurodegenerative diseases, inflammation biomarkers and detection of nucleic acids are easily possible through plasmonic SPR/SERS sensors.

5.4.1 Point of Care Diagnostic Devices (POCs)

Plasmonic biosensing is the most versatile analytical tool. By selecting the appropriate recognition element, it can be used for the detection of any type of target molecules as described in the previous sections. Medical/clinical purpose of plasmonic biosensors are most widely discussed. As POCs, plasmonic biosensors are the most interesting research topic in the biomedical field. POCs are highly relevant in the present scenario. Consistent monitoring of cholesterol, glucose, and other biomarkers are essential to the early detection of lifestyle diseases, cancers, or bacterial/viral diseases. Regular monitoring/bedside monitoring without established clinical laboratories are possible through POCs in resource limited areas. SO the healthcare system demands more efficient, cost-effective and reliable point of care diagnostic devices to upgrade our public health. Plasmonic biosensors can offer reliable, consistent and feasible results on the detection of biomarkers or any type of target molecules [37].

The analytes or target groups can be captured from the body fluids (blood serum, saliva, sweat or urine) by immobilizing specific ligands or antibodies on the nanoparticle surface. Tags and probes are also introduced for the better detection of target groups. Surface of the sensing substrate designed for plasmonic biosensors and surface chemistry and surface area are need to be considered. Clinical detection of biomarkers mainly focused on the concentration range of the analyte. The plasmonic biosensor should be detect or quantified the analytes in the desired physiological concentration range. Table 5.1 shows the application of plasmonic biosensors.

Table 5.1 Applications of plasmonic biosensors

Analyte	Functionalization	Detection method	Material/substrate	Sensitivity	Ref
Cytokines	Cytokine antibodies conjugation	LSPR/ microfluidics	Au nanorods microarray	5 pg mL^{-1}	[39]
MicroRNA	Hybridization between complementary probes	LSPR	Gold nano prisms attached to silanized glass	$\sim 10^{-13}$ M	[40]
Low molecular weight molecule (glycerol)	None	SPR	Hyperbolic metamaterials: 16 alternating thin films of gold and aluminium dioxide (Al_2O_3)	Ultra-low molecular mass of 266 Da	[10]
Thrombin	Thiol-modified thrombin aptamer was attached onto AuNPs/Fe$_3$O$_4$	LSPR/ absorption	AuNPs loaded on magnetic Fe$_3$O$_4$ nanoparticles	200 pM	[41]

Due to the versatility of plasmonic biosensors it can contribute to the detection of novel corona virus. It is feasible to fabricate the sensors for rapid diagnosis of COVID-19 viruses. It will be possible by detecting the genomic RNA sequences of the viruses. It is reported that the direct and label free detection of gene fragments of virus by immobilizing on the sensor substrate. Some reports are available about the contribution of plasmonic nanoparticles for the detection of corona virus. Guangyu et al. reported that the sensitive detection of SARS-CoV-2 using two-dimensional Au nano islands functionalized with complimentary DNA receptors [38]. The plasmonic biosensor was fabricated with the combination of plasmonic photothermal effect and LSPR sensing and leads to the potential diagnosis of COVID-19 virus. The localized heat generated through plasmonic photothermal effect from the Au nano islands is capable for the better performance of virus detection [38].

Nano plasmonic biosensors are used in other fields including environmental monitoring, food safety etc. Some of the important areas are described below.

5.4.2 Environmental Monitoring

The importance of environmental monitoring is the need of the day owing to the growing issues related to air and water pollution. Biosensors for environment monitoring mainly focus on the detection of trace amounts of heavy metals, pesticides, pathogens, and other chemicals in water and food and it can be used to detect the pollutants in a reliable and direct manner. Biosensors which are easily accessible and capable of rapid measurement of samples, are required for the detection of pollutants of different kind which we come across in our daily life. For the performance improvement of substrate in plasmonic bio sensing, the pollutants must be transferred as close as possible to the surface of the plasmonic active part. This can be achieved by functionalizing the substrates with specific recognition elements or by designing structures to allow for the easy binding of analytes. Also, plasmonic materials with higher affinity to analytes should be used. SERS biosensors based on Au NPs have been used for the detection of pesticides because of the higher affinity of Au NPs to dithiol carbamate. Toxic chemicals, toxic gases, pesticides, and heavy metal ion contents are the target analytes in the plasmonic biosensors in environmental monitoring [42, 43].

5.4.3 Food Safety

Food safety monitoring is highly important in daily life since the access to safe and nutritious food is the key for good health. Unsafe food can contain high levels of contaminants or adulterants like heavy metals, Food additives etc. which can cause serious health issues in humans [44, 45]. Plasmonic biosensors will help to detect the chemical and microbial contaminants in the food and these sensors must have features like high sensitivity, rapid detection, and simple preparation [43, 46]. Small target analytes are detected by the engineered substrates with specific binding sites. At the same time, for larger analyte molecules such as protein, virus or bacteria, the functionalisation of substrates with specific antibodies are needed. The sensitivity of detection of these types of analytes are much smaller. Toxic contaminants, preservatives, bacteria, heavy metal ion contents, food colour components and other toxic chemicals are the analytes in plasmonic sensors for food safety. X. Li et al. reported that melamine content in milk was detected using silver dendrite-based SERS biosensors with the sensitivity of 7.9×10^{-7} M [47]. It is an easy fabrication method with low cost. It is also reported that methylene blue content in fish was detected using SERS based Ag NPs sensor. This plasmonic biosensor was fabricated with Ag nanoflowers sandwiched between PMMA and graphene layers [48].

5.4.4 Substrate Characterization

There are numerous studies conducted on the Characterization of plasmonic substrates and materials before employing them for specific applications. Such studies give an indication about the overall performance and makes it possible to optimize further for real applications. Based on the relevant field of application, the analytes can contain DNA, proteins, viruses, and other biological particles like exosomes [49, 50]. Certain test molecules are commonly used for the characterization of plasmonic substrates in detection methods. Test molecules help in performance characterization of the substrate in terms of sensitivity or enhancement factor. The substrate performance is compared to Standard Raman measurements to illustrate the extent of surface enhancement due to the designed plasmonic nanostructures. In the case of characterization of systems which is highly sensitive to molecular interactions on the surface of the sensing substrate like SPR system, researchers mainly focus on the characterization of binding events [51]. Antigen–antibody interactions, IgG examinations, or other types of specific molecular interaction are some of the examples for the binding events. Analytes and methods of identification should be considered in the substrate development process.

5.5 Conclusion

Surface plasmon resonance is an excellent property that can be exploited for label free, efficient rapid and real time sensing of physiologically important analytes in biologically relevant concentration range, often very feeble to the range of nm. Due to their high surface enhancement of LSPR signal, detection of biologically relevant small molecules at very small concentrations can also be made possible which can help in the fabrication of portable, self-standing and compact point of care sensing devices for which exploration and fabrication of novel materials and combinations are essential.

References

1. J. Chao, W. Cao, S. Su et al., Nanostructure-based surface-enhanced Raman scattering biosensors for nucleic acids and proteins. J. Mater. Chem. B **4**, 1757–1769 (2016). https://doi.org/10.1039/C5TB02135A
2. J. Zheng, L. He (2014) Surface-enhanced Raman spectroscopy for the chemical analysis of food. Compr. Rev. Food Sci. Food Saf. **13**, 317–328. https://doi.org/10.1111/1541-4337.12062
3. P. Mehrotra, Biosensors and their applications–A review. J. oral. Biol. Craniofacial. Res. **6**, 153–159 (2016)
4. C.I.L. Justino, A.C. Duarte, T.A.P. Rocha-Santos, Critical overview on the application of sensors and biosensors for clinical analysis. TrAC Trends Anal Chem **85**, 36–60 (2016)
5. C.I.L. Justino, A.C. Duarte, T.A.P. Rocha-Santos, Recent progress in biosensors for environmental monitoring: A review. Sensors **17**, 2918 (2017)
6. J. Liu, M. Jalali, S. Mahshid, S. Wachsmann-Hogiu, Are plasmonic optical biosensors ready for use in point-of-need applications? Analyst **145**, 364–384 (2020). https://doi.org/10.1039/c9an02149c
7. X. Michalet, F.F. Pinaud, L.A. Bentolila, J.M. Tsay, S. Doose, J.J. Li et al, Quantum dots for live cells. Vivo Imaging, Diagnostics Sci. **307**, 538–544 (2005)
8. C.M. Tyrakowski, P.T. Snee, A primer on the synthesis, water-solubilization, and functionalization of quantum dots, their use as biological sensing agents, and present status. Phys. Chem. **16**, 837–855 (2014)
9. M.E. Stewart, C.R. Anderton, L.B. Thompson et al., Nanostructured plasmonic sensors. Chem. Rev. **108**, 494–521 (2008)
10. K.V. Sreekanth, Y. Alapan, M. ElKabbash et al., Extreme sensitivity biosensing platform based on hyperbolic metamaterials. Nat. Mater. **15**, 621–627 (2016)
11. J. Lee, K. Takemura, E.Y. Park, Plasmonic nanomaterial-based optical biosensing platforms for virus detection. Sensors **17**, 2332 (2017)
12. S.D. Soelberg, R.C. Stevens, A.P. Limaye, C.E. Furlong, Surface plasmon resonance detection using antibody-linked magnetic nanoparticles for analyte capture, purification, concentration, and signal amplification. Anal. Chem. **81**, 2357–2363 (2009)
13. M. Svedendahl, R. Verre, M. Käll, Refractometric biosensing based on optical phase flips in sparse and short-range-ordered nanoplasmonic layers. Light. Sci. Appl. **3**, e220–e220 (2014)
14. J. Jatschka, A. Dathe, A. Csáki et al., Propagating and localized surface plasmon resonance sensing—A critical comparison based on measurements and theory. Sens. Bio-sensing Res. **7**, 62–70 (2016)
15. K.A. Willets, R.P. Van Duyne, Localized surface plasmon resonance spectroscopy and sensing. Annu. Rev. Phys. Chem. **58**, 267–297 (2007)
16. A.J. Haes, C.L. Haynes, A.D. McFarland et al., Plasmonic materials for surface-enhanced sensing and spectroscopy. MRS Bull. **30**, 368–375 (2005). https://doi.org/10.1557/mrs2005.100
17. Ashcroft NW, Mermin ND (1976) Solid state physics Brooks. Cole, Cengage Learn 10:
18. E. Petryayeva, U.J. Krull, Localized surface plasmon resonance: Nanostructures, bioassays and biosensing—A review. Anal. Chim. Acta. **706**, 8–24 (2011). https://doi.org/10.1016/j.aca.2011.08.020
19. A.J. Haes, L. Chang, W.L. Klein, R.P. Van Duyne, Detection of a biomarker for Alzheimer's disease from synthetic and clinical samples using a nanoscale optical biosensor. J. Am. Chem. Soc. **127**, 2264–2271 (2005)

20. J. Kim, S.Y. Oh, S. Shukla et al., Heteroassembled gold nanoparticles with sandwich-immunoassay LSPR chip format for rapid and sensitive detection of hepatitis B virus surface antigen (HBsAg). Biosens. Bioelectron. **107**, 118–122 (2018)
21. E. Mauriz, Recent progress in plasmonic biosensing schemes for virus detection. Sensors **20**, 4745 (2020)
22. T. Lee, G.H. Kim, S.M. Kim et al. (2019) Label-free localized surface plasmon resonance biosensor composed of multi-functional DNA 3 way junction on hollow Au spike-like nanoparticles (HAuSN) for avian influenza virus detection. Colloids Surf. B Biointer. **182**, 110341
23. N. Bellassai, R. D'Agata, V. Jungbluth, G. Spoto, Surface plasmon resonance for biomarker detection: Advances in Non-invasive cancer diagnosis. Front. Chem. **7**, 570 (2019). https://doi.org/10.3389/fchem.2019.00570
24. W. Zhou, Y. Ma, H. Yang et al., A label-free biosensor based on silver nanoparticles array for clinical detection of serum p53 in head and neck squamous cell carcinoma. Int. J. Nanomedicine. **6**, 381 (2011)
25. W.S. Hwang, S.J. Sim, A strategy for the ultrasensitive detection of cancer biomarkers based on the LSPR response of a single AuNP. J. Nanosci. Nanotechnol. **11**, 5651–5656 (2011). https://doi.org/10.1166/jnn.2011.4346
26. T. Huang, P.D. Nallathamby, X.-H.N. Xu, Photostable single-molecule nanoparticle optical biosensors for real-time sensing of single cytokine molecules and their binding reactions. J. Am. Chem. Soc. **130**, 17095–17105 (2008)
27. X. Ma, K. Fletcher, T. Kipp et al., Photoluminescence of individual Au/CdSe nanocrystal complexes with variable interparticle distances. J. Phys. Chem. Lett. **2**, 2466–2471 (2011)
28. C. Xue, Y. Xue, L. Dai et al., Size-and shape-dependent fluorescence quenching of gold nanoparticles on perylene dye. Adv. Opt. Mater. **1**, 581–587 (2013)
29. J.R. Lakowicz, K. Ray, M. Chowdhury et al., Plasmon-controlled fluorescence: a new paradigm in fluorescence spectroscopy. Analyst **133**, 1308–1346 (2008)
30. N. De Acha, C. Elosua, I. Matias, F.J. Arregui, Luminescence-based optical sensors fabricated by means of the layer-by-layer nano-assembly technique. Sensors **17**, 2826 (2017)
31. M. Bauch, K. Toma, M. Toma et al., Plasmon-enhanced fluorescence biosensors: a review. Plasmonics **9**, 781–799 (2014)
32. L. Wang, Q. Song, Q. Liu et al. (2015) Plasmon-enhanced fluorescence-based core–shell gold nanorods as a near-IR fluorescent turn-on sensor for the highly sensitive detection of pyrophosphate in aqueous solution. Adv. Funct. Mater. 25, 7017–7027. https://doi.org/10.1002/adfm.201503326
33. K.A. Willets, Super-resolution imaging of interactions between molecules and plasmonic nanostructures. Phys. Chem **15**, 5345–5354 (2013)
34. H. Cang, A. Labno, C. Lu et al., Probing the electromagnetic field of a 15-nanometre hotspot by single molecule imaging. Nature **469**, 385–388 (2011)
35. M. Thomas, R. Carminati, J.-J. Greffet, J.R. Arias-Gonzalez JR, Single molecule spontaneous emission close to absorbing metallic nanostructures (2004)
36. K. Takemura, O. Adegoke, N. Takahashi et al., Versatility of a localized surface plasmon resonance-based gold nanoparticle-alloyed quantum dot nanobiosensor for immunofluorescence detection of viruses. Biosens. Bioelectron. **89**, 998–1005 (2017)
37. A.M. Shrivastav, U. Cvelbar, I. Abdulhalim, A comprehensive review on plasmonic-based biosensors used in viral diagnostics. Commun. Biol. **4**, 1–12 (2021)
38. G. Qiu, Z. Gai, Y. Tao et al., Dual-functional plasmonic photothermal biosensors for highly accurate severe acute respiratory syndrome coronavirus 2 detection. ACS Nano. **14**, 5268–5277 (2020)
39. P. Chen, M.T. Chung, W. McHugh et al., Multiplex serum cytokine immunoassay using nanoplasmonic biosensor microarrays. ACS Nano. **9**, 4173–4181 (2015)
40. G.K. Joshi, S. Deitz-McElyea, M. Johnson et al., Highly specific plasmonic biosensors for ultrasensitive microRNA detection in plasma from pancreatic cancer patients. Nano. Lett. **14**, 6955–6963 (2014)

41. J. Yan, L. Wang, L. Tang et al., Enzyme-guided plasmonic biosensor based on dual-functional nanohybrid for sensitive detection of thrombin. Biosens. Bioelectron. **70**, 404–410 (2015)
42. N. Chen, P. Ding, Y. Shi et al., Portable and reliable surface-enhanced Raman scattering silicon chip for signal-on detection of trace trinitrotoluene explosive in real systems. Anal. Chem. **89**, 5072–5078 (2017)
43. J. Homola, Present and future of surface plasmon resonance biosensors. Anal. Bioanal. Chem. **377**, 528–539 (2003)
44. E.Y.Y. Chan, S.M. Griffiths, C.W. Chan, Public-health risks of melamine in milk products. Lancet **372**, 1444–1445 (2008)
45. R. Najafi, S. Mukherjee, J. Hudson Jr. et al., Development of a rapid capture-cum-detection method for Escherichia coli O157 from apple juice comprising nano-immunomagnetic separation in tandem with surface enhanced Raman scattering. Int. J. Food Microbiol. **189**, 89–97 (2014)
46. J. Chen, Y. Huang, P. Kannan et al., Flexible and adhesive surface enhance Raman scattering active tape for rapid detection of pesticide residues in fruits and vegetables. Anal. Chem. **88**, 2149–2155 (2016)
47. X. Li, S. Feng, Y. Hu et al., Rapid detection of melamine in milk using immunological separation and surface enhanced Raman spectroscopy. J. Food Sci. **80**, C1196–C1201 (2015)
48. H. Qiu, M. Wang, S. Jiang et al., Reliable molecular trace-detection based on flexible SERS substrate of graphene/Ag-nanoflowers/PMMA. Sensors Actuators B Chem. **249**, 439–450 (2017)
49. M. Kahraman, E.R. Mullen, A. Korkmaz, S. Wachsmann-Hogiu, Fundamentals and applications of SERS-based bioanalytical sensing. Nanophotonics **6**, 831–852 (2017)
50. C. Wang, F. Madiyar, C. Yu, J. Li, Detection of extremely low concentration waterborne pathogen using a multiplexing self-referencing SERS microfluidic biosensor. J. Biol. Eng. **11**, 1–11 (2017)
51. H.H. Nguyen, J. Park, S. Kang, M. Kim, Surface plasmon resonance: a versatile technique for biosensor applications. Sensors **15**, 10481–10510 (2015)

Chapter 6
Nonlinear Optical Properties of Nanomaterials

Pranitha Sankar and Reji Philip

6.1 Introduction

Nonlinear optics (NLO) involves the study of the interaction of intense light fields with matter. After the invention of the laser in 1960, Franken and colleagues [1] observed second harmonic generation (SHG) of ruby laser beam in a quartz crystal, which was followed by similar studies in various materials. Thereafter, Bloembergen and colleagues formulated a general theoretical framework for three- and four-wave mixing at optical frequencies [2]. The quantum mechanical calculation of complex nonlinear susceptibilities, based on the evolution of the density matrix (see for example, [3, 4]), was subsequently applied to optical problems. In 1965 Rentzepis and Pao reported the first observation of SHG in an organic material (benzpyrene) [5].

The field of NLO grew substantially over the years, facilitating a deeper understanding of light-matter interaction and providing solutions for several technological problems. Efficient nonlinear optical interactions have become essential for many applications in advanced photonics. However, they typically require intense laser sources, different wavelengths and interaction lengths, in order to utilize nonlinear optics for novel nanophotonic architectures in integrated optics and metasurface devices. Obtaining materials with stronger nonlinear properties is an important step towards applications. For example, certain essential devices in the telecom industry such as switches, routers, and wavelength converters can be realized using optical nonlinearity. While second order nonlinearities mostly involve optical frequency

P. Sankar
Institut Für Quantenoptik, Leibniz Universität Hannover, 30167 Hannover, Germany

R. Philip (✉)
Light and Matter Physics Group, Raman Research Institute, C. V. Raman Avenue, Sadashivanagar, Bangalore 560080, India
e-mail: reji@rri.res.in

conversion, third order phenomena include nonlinear phase modulation, nonlinear absorption and nonlinear scattering of light. Nanoparticles and nanocomposites have become important materials for nonlinear optics in recent years because of their interesting surface effects and properties such as quantum confinement, plasmon oscillations etc., which are absent in the corresponding bulk materials. The remarkable NLO properties displayed by nanomaterials have motivated the design and fabrication of nanoscale photonic and optoelectronic devices [6]. Several practical optical devices using 2-D layered materials, such as optical modulators, optical polarizers, optical switchers, and even all-optical devices, are expected to be developed in the near future [7].

Carbon-based nanomaterials such as 0D fullerenes, 1D carbon nanotubes (CNT) and 2D graphenes [8] have become essential materials for nanotechnology, and their spectacular mechanical, electrical and thermal properties, in addition to their distinctive NLO properties, have generated substantial research interest from both academic and industrial aspects [9–11]. Metamaterials are another class of engineered materials made from meta-atom building blocks organized into arrays of subwavelength metal and/or dielectric components. Strong light-matter interaction in metamaterials results in novel linear and nonlinear optical (NLO) properties [12], making them attractive for applications in sensing, terahertz imaging, and optical fiber manufacture. Recently, it has been established that materials with a vanishingly small permittivity can enable efficient nonlinear optical phenomena [13]. These materials are commonly known as epsilon-near-zero (ENZ) materials, and they exhibit unprecedented ultrafast nonlinear efficiencies within sub-wavelength propagation lengths with low optical losses and high nonlinear enhancements [14].

In this chapter we will outline the physical origins of optical nonlinearity in material media, and discuss the nonlinear optical properties reported for advanced nanomaterials in recent years. Experimental techniques will be explained and results obtained in different nanomaterials will be overviewed.

6.2 The Nonlinear Susceptibility

A steady electric field $\mathbf{E}$ applied to a dielectric medium generates an ensemble of induced dipoles. The net average dipole moment per unit volume (electric polarization) is given by

$$\mathbf{P} = N<\mu> \tag{6.1}$$

where N is the number of microscopic dipoles per unit volume, and $<\mu>$ is the ensemble averaged induced dipole moment. $\mathbf{P}$ can also be written in the form

$$\mathbf{P_i} = \varepsilon_0 \chi_{ij}^{(1)} \mathbf{E_j} \tag{6.2}$$

where ε_0 is the vacuum dielectric constant, $\chi_{ij}^{(1)}$ w is the linear susceptibility, and $\mathbf{E_j}$ is the amplitude of the electric field vector. $\chi_{ij}^{(1)}$ is a second rank tensor since it relates the two vector quantities $\mathbf{P_i}$ and $\mathbf{E_j}$. Now, it is possible to consider the electric field of electromagnetic radiation in the place of a steady electric field, in which case $\mathbf{E_j}$ and $\mathbf{P_i}$ become rapidly oscillating vector fields. The linear optical properties of the medium, viz. the refractive index (n_0) and absorption coefficient (α_0), are then derived from $\chi_{ij}^{(1)}$.

To interpret the observed effects of second order and third order nonlinearities in different materials, it is necessary to speculate that under the action of an intense laser, the dielectric strength of the medium is no longer linearly related to the amplitude of the incident light field. If the intensity of the exciting electromagnetic radiation is sufficiently strong, the dipole oscillations become anharmonic. The polarization will have a nonlinear dependence on the electric field in this case [4], which can be written as a power series expansion

$$\mathbf{P_i} = \varepsilon_0 \left\{ \chi_{ij}^{(1)} \cdot \mathbf{E_j} + \chi_{ijk}^{(2)} \cdot \mathbf{E_j}\mathbf{E_k} + \chi_{ijkl}^{(3)} \cdot \mathbf{E_j}\mathbf{E_k}\mathbf{E_l} + \cdots \right\} \qquad (6.3)$$

where $\chi_{ijk}^{(2)}$ and $\chi_{ijkl}^{(3)}$ are the second order and third order nonlinear optical suscepti-bilities respectively. In general, the dipole response to an applied field can always be considered as anharmonic, and therefore Eq. (6.3) becomes the generalized equation for the response of a medium to electromagnetic radiation.

The magnitude of the optical nonlinearity will depend on the applied light field intensity and the nonlinear susceptibility coefficients. $\chi^{(2)}$ leads to second order nonlinearities and $\chi^{(3)}$ leads to third order nonlinearities in the medium. Media with inversion symmetry do not exhibit nonlinearities of even order, while those without a symmetry axis display nonlinearity of both even and odd orders. Therefore, while the lowest nonlinearity in anisotropic crystals is of the second order, that in isotropic media like liquids, gases and amorphous solids (e.g.: glass) is of the third order.

While second order nonlinearities in crystals mostly lead to frequency conver-sion phenomena such as harmonic generation and parametric mixing, the third order nonlinear susceptibility $\chi^{(3)}$ contributes to nonlinear refraction, nonlinear absorp-tion, and nonlinear scattering (stimulated Raman and stimulated Brillouin scattering) effects.

6.3 Nonlinear Refraction and Absorption

Nonlinear refraction occurs when there is a change in the refractive index of a medium due to the presence of external fields. A number of physical effects such as optical Kerr effect, thermal lensing, electrostriction and population redistribution can contribute to nonlinear index of refraction. It may be noted that some of these like thermal lensing are resonant phenomena and the calculation of the corresponding susceptibilities requires a treatment which is different from the perturbative approach adopted in Eq. 6.3.

Considering the optical Kerr effect, the real and imaginary parts of the corresponding degenerate nonlinear susceptibility, $\chi^{(3)}$ ($\omega; \omega, -\omega, \omega$) are related to self-focusing and two-photon absorption. These are given by [15]

$$\text{Re}\,\chi^{(3)} = (4/3)n_0^2 n_2 \varepsilon_0 c \tag{6.4}$$

and

$$\text{Im}\,\chi^{(3)} = (\lambda/3\pi)n_0^2 \alpha_2 \varepsilon_0 c \tag{6.5}$$

where λ is the light wavelength in meters, c is the light velocity in ms^{-1}, α_2 is the two-photon absorption coefficient in mW^{-1}, and n_2 is the nonlinear refractive index in $m^2 W^{-1}$. The net refractive index of the medium is given by $n = n_0 + n_2 I$, where I is the light intensity in Wm^{-2}. I is related to the optical field E through the relation

$$I = 2n_0(\varepsilon_0/\mu_0)^{1/2}|E|^2 = (2n_0/Z_0)|E|^2 \tag{6.6}$$

where $\varepsilon_0 = 8.85 \times 10^{-12}\,Fm^{-1}$ is the permittivity, $\mu_0 = 4\pi \times 10^{-7}$ H/m is the permeability, and $Z_0 = 377\,\Omega$ is the characteristic impedance, of free space. Depending on whether n_2 is positive or negative, the modification in the refractive index will give rise to self-focusing or defocusing effects. Similarly, the net absorption coefficient is given by $\alpha = \alpha_0 + \alpha_2 I$, so that a modified Beer-Lambert law can be written in the form

$$I = I_0 e^{-(\alpha_0 + \alpha_2 I)z} \tag{6.7}$$

where z is the direction of propagation. A positive α_2 will lead to enhanced absorption while a negative α_2 will lead to enhanced transmission.

Nonlinear refraction has potential applications in optical switching, optical limiting, passive laser mode-locking, and waveguide switches and modulators. At low light levels the absorption coefficient of a medium does not depend on of the input light intensity. At high intensities, however, several phenomena can take place that can modify this behavior. For instance, absorption can get saturated as the excited state population increases. Alternatively, the excited states can absorb photons, sometimes even stronger than the ground state. Similarly, photo-generated free carriers in semiconductors may absorb more photons. Another possibility is that of two or more photons getting absorbed in a single event. All these phenomena will change the transmittance of the medium at high intensities, causing a deviation from the Beer-Lambert law. This change in transmittance of a material with input light intensity or fluence is generally known as nonlinear absorption. A few of these processes are saturable absorption, excited state absorption, multi-photon absorption and free-carrier absorption.

6.4 Measurement of Second Order Nonlinearities

Second-order nonlinear optical (NLO) materials have attracted much attention primarily because of their excellent ability for wavelength conversion. Experimental methods in second order NLO are usually devised to measure either the nonlinear coefficient d_{eff} or the first order hyperpolarizability β. Two absolute measurement methods often used are the phase-matched method and parametric fluorescence method respectively. Maker fringe method [16] and Kurtz powder method [17] are the relative measurement techniques employed for characterizing the medium with respect to a standard reference material. These techniques are useful when the sample is in single crystal form. For thin films reflectance measurements can be done, and the poling technique [15] is found to be useful for polymer thin films.

While developing new materials for NLO it may often be convenient to characterize individual molecules initially, before proceeding to study the bulk material. This is particularly true for organic media since bulk properties are largely determined by the individual molecular units. Relative methods for characterizing molecular parameters include Electric field induced second harmonic generation (EFISH) [18] which is useful in the case of neutral, dipolar molecules, and Hyper-Rayleigh Scattering (HRS) [19] for non-dipolar or charged molecules. Various theoretical approaches have been developed for the calculation of the molecular hyperpolarizability [20]. Even though the above methods are most useful approaches for chemists to understand structure–hyperpolarizability relationships, and for designing new, highly efficient second-order NLO molecular materials.

6.5 Measurement of Third Order Nonlinearities

Experimental methods for third order nonlinearity determination include Third harmonic generation, degenerate four wave mixing, Z-scan, optical Kerr effect, and two-photon fluorescence. Of these, the Z-scan is the simplest technique to implement. Within a short period of its invention the Z-scan has become a major method for measuring the nonlinear refraction and absorption coefficients in materials.

Z-scan: The nonlinear absorption coefficient and nonlinear refractive index of a material can be measured using the open aperture and closed aperture Z-scan techniques respectively [21]. In the open aperture configuration, a laser beam having a Gaussian spatial distribution is initially focused by a lens. The beam direction is taken as the z axis, and the beam focus is taken as $z = 0$. The sample to be measured is then translated in short steps from a negative z position to a positive z position through the focal point. The sample transmission corresponding to each z position is

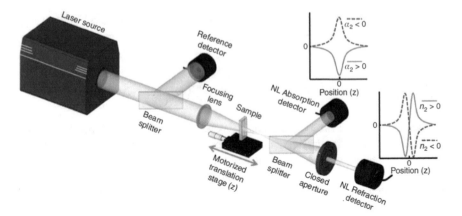

Fig. 6.1 Schematic of a simultaneous open aperture and closed aperture Z-scan experiment. The NL absorption detector measures open aperture signal while the NL refraction detector measures closed aperture signal, as a function of sample position (from Ref. [12])

measured using a detector placed after the sample. A graph in which the measured transmission (usually normalized to the linear transmission value) is plotted against z is known as the Z-scan curve. In the closed aperture Z-scan a small aperture is placed in front of the detector, which makes the measurement sensitive to beam size variations due to nonlinear refraction. The Z-scan experimental set up is shown in Fig. 6.1.

Open aperture Z-scan: In open aperture Z-scan no aperture will be kept in front of the detector, and therefore, the detector will measure all of the light transmitted by the sample. When the sample is at the focal point its transmission will be either a maximum or minimum, depending on the sign of the dominant nonlinear mechanism. For reverse saturable absorption the transmittance will be a minimum, and for saturable absorption it will be a maximum. In general, an open-aperture Z-scan trace will be symmetric with respect to the focus ($z = 0$). The normalized transmittance of a laser pulse through a third order nonlinear medium is given by [21]

$$T(z) = \left[\frac{1}{\pi^{\frac{1}{2}} q(z)} \right] \int_{-\infty}^{+\infty} \ln\left[1 + q(z)\exp\left(-\tau^2\right)\right] d\tau \qquad (6.8)$$

with $q(z) = \alpha_2 I_0 L_{\text{eff}}/[1 + (z/z_0)^2]$, where I_0 is the peak intensity at the focal point. α_2 denotes the nonlinear absorption coefficient. For a medium that is transparent at the excitation wavelength, α_2 will be the two-photon absorption coefficient. The magnitude and sign of α_2 can be determined by numerically fitting the open aperture Z-scan curve to Eq. 6.8. Expressions for Z-scan transmittance under multiphoton excitation conditions are available in literature [22, 23].

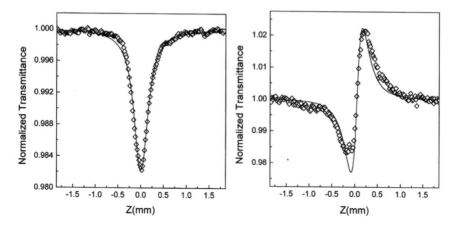

Fig. 6.2 Open aperture and closed aperture Z-scan curves measured in Ag nanorods dispersed in borosilicate glass, using 800 nm, 240 fs laser pulses at an intensity of 8.8×10^9 W/cm^2. Numerical fits (solid curves) to the experimental data (squares) reveal a third order nonlinearity, giving an α_2 value of 1.73×10^{-9} cm W^{-1} and n_2 value of 2.0×10^4 cm^2/GW (From Ref. [89])

A typical open-aperture Z-scan curve, measured in Ag nanorods dispersed in borosilicate glass, is shown in Fig. 6.2. The figure shows a "valley" which is typical for samples showing two-photon absorption or excited state absorption. In this case, the presence of the strong surface plasmon resonance (SPR) band around 400 nm facilitates strong two-photon absorption when the nanorods are excited at 800 nm.

Closed aperture Z-scan: The closed aperture Z-scan measurement is based on the self-focusing or self-defocusing of a spatially Gaussian optical beam occurring in a thin nonlinear medium. A plot of normalized transmittance versus sample position in this configuration will give information regarding the real part of the nonlinearity (nonlinear refractive index). As mentioned earlier, the net refractive index of a medium with a Kerr nonlinearity can be written as $n = n_0 + n_2 I$, where n_0 is the linear refractive index, n_2 is the nonlinear refractive index coefficient, and I is the time averaged intensity of the optical field. Now consider a laser beam with a Gaussian beam cross section, given by $I = I_0 \exp(-2r^2/\omega^2)$, where ω is the beam radius and r is the radial co-ordinate. The intensity of the beam will be a maximum at the center, and will drop along the radial direction. Therefore, when a Gaussian laser beam passes through a nonlinear medium its refractive index will be modified in such a way that maximum change occurs at the beam center. This results in a phase modulation of the wavefront such that the beam gets focused if n_2 is positive, and de-focused if n_2 is negative. Thus the medium will behave as a nonlinear lens, the focal length of which is intensity-dependent, and hence depends on the sample position z. Therefore, if a small aperture is placed in front of the detector, then the aperture transmittance will become a function of z. Such focusing/defocusing is usually small in magnitude, and a sufficiently high intensity (I), as from a pre-focused beam, is normally required to get a measurable effect.

A typical closed aperture Z-scan measured in Ag nanorods at the excitation wavelength of 800 nm is shown in Fig. 6.2. The figure shows a "valley and peak" structure. This symmetric "valley-peak" structure is typical for a sample that shows self-focusing (a "peak-valley" structure will appear for a self-defocusing sample). The difference in normalized transmittance between the peak and the valley, ΔT_{p-v}, is a measure of the nonlinear refractive index of the medium. From numerical fits to theory, the nonlinear refractive index of the Ag nanorods is calculated to be $n_2 = 2.0 \times 10^4$ cm^2/GW.

Considering a thin nonlinear medium (i.e., $l \ll z_R$ where l is the sample length and $z_R = \pi \omega_0^2 / \lambda$ is the Rayleigh range of the focused beam in air, where ω_0 is the $1/e^2$ beam radius at focus and λ is the excitation wavelength), exhibiting the optical Kerr effect and having no nonlinear absorption, the nonlinear phase $\Delta \phi(z, t)$ impressed on the wave is given by

$$\Delta \phi(z, t) = \Delta \phi_0(t) / \left[1 + (z/z_R)^2 \right] \tag{6.9}$$

When the sample is at the beam focus the phase distortion will be a maximum. If the on-axis phase shift is small (i.e., when $|\Delta \phi_0| < \pi$), then ΔT_{p-v} is approximately (within a precision of 3%) given by

$$\Delta T_{p-v} = 0.405(1 - S)^{0.25} {<} \Delta \phi_0 {>} \tag{6.10}$$

where S is the linear aperture transmittance given by $1 - \exp(-2r_a^2/\omega_a^2)$. Here r_a is the aperture radius and ω_a is the beam radius at the aperture. Knowing ΔT_{p-v} from the z-scan curve, ${<} \Delta \phi_0 {>}$ can be estimated using the above equation. The nonlinear refractive index coefficient can then be determined from the relation

$$\Delta T_{p-v} = 0.405(1 - S)^{0.25} {<} \Delta \phi_0 {>} \tag{6.11}$$

where E_i is the input energy, t_{FWHM} is the laser pulse width (full width at half maximum), and $L_{eff} = [1 - \exp(-\alpha_0 l)]/\alpha_0$.

If the on-axis phase shift is larger than π (like in the case of a strong thermal lens effect), then the above approximations have to be replaced by more precise expressions. Similarly, if the sample shows both nonlinear absorption and nonlinear refraction simultaneously, then the closed-aperture curve should be divided by the open aperture curve before carrying out the nonlinear refractive index calculations.

6.6   Enhancement of Optical Nonlinearities in Nanomaterials

In this section, we focus on the nonlinear optical response of plasmonic and carbon-based nanomaterials or their composites, which ideally possess NLO properties superior to those of the starting materials. These are classified into: '0' dimension materials like nanospheres and nanocrystals, '1' dimension materials like nanotubes and nanorods, and '2' dimension materials like nanofilms, nanofoils and nano surface layers. The usually encountered structures of nanomaterials can be classified into four different major groups, namely the Maxwell-Garnett geometry, Bruggeman geometry, layered structures, and fractal structures.

6.6.1   Maxwell-Garnett Geometry

In this geometry, the nanoparticle is dispersed randomly in a host material as shown in Fig. 6.3. According to this model, the particles are of uniform size and spherical shape. Optical properties of composite materials in this geometry can be understood in terms of local field effects. The concept of local fields [24] is important in standard discussions of the optical properties of homogenous nanomaterials. For example, in the presence of an oscillating electric field, metal nanospheres emit electric dipole radiation. Maxwell-Garnett theory replaces the spheres in the model by the equivalent point dipoles, i.e., their finite size is ignored. If $\mathbf{p}$ represents the average dipole moment of an inclusion and N represents the number of nanoparticles per unit volume, then the total polarization of the medium (normalized by the dielectric constant of the host) is given by,

$$\mathbf{P} = N\mathbf{p} \tag{6.12}$$

The average dipole moment is given by,

$$p = a^3 \frac{\epsilon_i - \epsilon_h}{\epsilon_i + 2\,\epsilon_h} E_{loc} \tag{6.13}$$

Fig. 6.3 Maxwell-Garnett geometry (From Ref. [24])

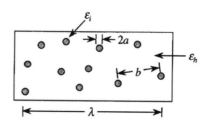

where 'a' is the nanoparticle radius and E_{loc} is the local field experienced by the nanoparticle. For sparse, randomly distributed dipoles, the local field is given by the Lorentz relation:

$$E_{loc} = E_0 + \frac{4\pi}{3\,\epsilon_h}P \qquad (6.14)$$

Although only metallic particles are considered, the derivation is valid in general, i.e., the constituents may be pure dielectrics as well. The above mentioned local electric field value depends on the polarization of the surrounding medium and also on the applied field, and it is not the same as the macroscopic electric field appearing in Maxwell's equations. Local-field effects play a role in each of the composite geometries to be considered, and is quite significant in nonlinear effective medium theories since the local-field correction factor appears multiple times in expressions for the nonlinear susceptibilities.

6.6.2 Bruggeman Geometry

This model assumes that grains of two or more materials are randomly interspersed (Fig. 6.4). The intermixed components possess different or enhanced linear and nonlinear optical properties. To analyze composites of Bruggeman geometry, one considers a single grain within the whole. Grains of each type of constituent material will surround this grain. The grains are considered to be surrounded by a material of uniform dielectric constant given by that of the effective medium. If we take the grain to be spherical, we may solve for the internal electric field due to a uniform applied field E_0: The linear optical properties are described by the equation,

$$E_i = \frac{3\,\epsilon_{eff}}{\epsilon + 2\,\epsilon_{eff}}E_0 \qquad (6.15)$$

which leads to a displacement field of the form

$$D_i = \epsilon_i \frac{3\,\epsilon_{eff}}{\epsilon_i + 2\,\epsilon_{eff}}E_0 \qquad (6.16)$$

Fig. 6.4 Bruggeman (interspersed) geometry (From Ref. [24])

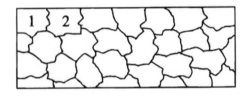

This theory will be applicable for materials in which all components occupy large volume fractions of the whole. An important feature of this model is that, as the volume fraction of one constituent increases from a small value, it will reach a point at which the grains begin to join and form continuous threads throughout the composite. At this stage, the Bruggeman theory accounts superbly over the Maxwell-Garnett model for the inclusion of conductivity of a metal/insulator composite and the behavior near the surface plasmon resonance of metal particles in a dielectric [25]. At very low volume fractions of the metallic constituent, the predictions of the two models are virtually identical. However, as the volume fraction increases the Maxwell-Garnett model continues to predict a sharp resonance, whereas the Bruggeman model predicts a broadening and weakening of the resonance.

6.6.3 Layered Geometry

Another composite geometry is that of alternating layers of two or more materials, shown in Fig. 6.5. Such a composite is considered to be anisotropic and uniaxial because the optical properties measured by using electric fields polarized parallel to the layers will be typically different from those for electric fields polarized perpendicular to the layers. To develop an effective medium theory, it is assumed that each layer is much thinner than an optical wavelength. For electric fields parallel to the layers the field is continuous across the boundaries of each layer, and the effective linear dielectric constant and nonlinear susceptibilities can be written as averages of the constituents:

$$\epsilon_{||} = f_1 \, \epsilon_1 + f_2 \, \epsilon_2 \qquad (6.17)$$

and

$$\chi^{(2)}_{sum} = f_1 \chi^{(2)}_1 + f_2 \chi^{(2)}_2 \qquad (6.18)$$

Here f represents the volume fill fraction of the nanoparticles or layers. When the electric field is polarized perpendicular to the layers, more interesting effects occur. In this case, the field is continuous across the layer boundaries, and the electric field is

Fig. 6.5 Layered geometry
(From Ref. [24])

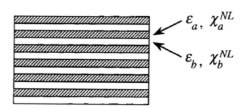

distributed non-uniformly between the layers, resulting in the nonlinear susceptibility becoming more complicated.

6.6.4   Fractal Structures

Another type of nanocomposite is formed when one of the constituents forms fractal structures within the whole. For example, one constituent could consist of metal spheres or dielectric nanoparticles that clump together forming aggregates with a fractal dimension. The mode in which the clusters form will affect their fractal dimension, which in turn affects the nonlinear optical properties. Since fractals do not possess translational symmetry, they cannot propagate pure traveling waves. However, local-field effects can be strong in fractal structures. Due to their geometry localized field excitations arise, which may lead to regions of enhanced absorption [26]. The scattering cross section is found to increase with the number of monomers until the size of the fractal becomes comparable to an optical wavelength, when saturation occurs [27]. Hui and Stroud analyzed the linear properties of fractal composites using a three-dimensional differential effective medium approach on self-similar clusters [28]. The approach of Shalaev and Stockman to fractal composites used a binary approximation, which considers the dipole-dipole coupling of the monomer to its nearest neighbor. The effect of the other monomers was found out by using a modified Lorentz local field [29].

Determination of the nonlinear optical properties of fractal composites is a more difficult task. Butenko et al. used the binary approximation to determine enhancement factors arising from the nonlinearities of impurities linked to the fractal for various nth-order nonlinear processes [30]. This factor could be very large due to the strong localized fields within the fractal structure. A number of other approximations also can be found in literature for the optical properties of fractal structures.

6.7   Optical Nonlinearity Measurements in Nanomaterials

Several authors have reported nonlinear optical properties of nanomaterials in literature, and a selected few are given below. Experimental results are broadly classified into second order and third order effects. The key requirements for practical nonlinear optical materials are fast response time, strong nonlinearity, broad wavelength range, low optical loss, high power handling, and ease of integration into an optical system. Since substantial advances have been made during the last 10 years in the design of plasmonic and carbon based nanomaterials, we focus on those materials below.

6.7.1 Second Order Nonlinearities

Second order nonlinearities occur in crystalline media with a non-centrosymmetric structure. NLO crystals such as potassium dihydrogen phosphate (KDP), lithium niobate ($LiNbO_3$), and super lattices grown by molecular-beam epitaxy (MBE) are readily available, but it is difficult to incorporate these materials into most photonic platforms. So, there is substantial current research interest on nonlinear optical nanomaterials for harmonic generation and frequency conversion applications.

Second harmonic generation (SHG) was first demonstrated by Franken et al. at the University of Michigan in 1961. They focused a ruby laser beam (694 nm) into a quartz crystal, and the output measured through a spectrometer and recorded on photographic paper revealed the production of light at 347 nm [31]. There is strong recent interest in the optical response of metal nanoparticles in view of the local field enhancements [32]. Giant local field enhancement factors of the order of 10^6 have been predicted, which are significant for nonlinear optical processes such as SHG and THG in nanomaterials.

An overview of how the second order NLO properties of nanomaterials vary with size, symmetry, and shape is given in the review by Ray [33]. In addition, the emergence of second order NLO nanomaterials for the development of nanomaterial based optical technology has also been discussed. The advancements in the design of nanomaterials of different sizes and shapes for NLO is highlighted. SHG from metal nanoparticles is typically attributed to electric dipole excitations at their surfaces caused by strong polarization, but nonlinearities involving higher multipole effects may also be significant due to strong nanoscale gradients in the local material properties and fields. The paper also summarizes recent advancements on the development of a nanomaterials based NLO assay for monitoring chemical processes, and sensing biomolecules and toxic metals [33] (Fig. 6.6).

Nonlinear plasmonic effects have applications including switching and modulation of optical signals, frequency conversion, and soliton generation. Tuning of plasmonic resonances in metamaterials and utilization of their properties will be important for the development of next-generation, low-power, all optical switches and all-optically-tunable plasmonic devices. Similarly, the advantages of GaAs nanowires

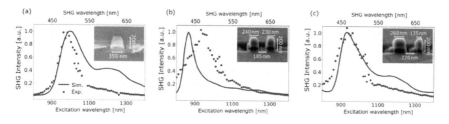

Fig. 6.6 Measured and calculated second harmonic emission of **a** single **b** symmetric dimer **c** asymmetric dimers of some III–IV semiconductor nanorods. SEM images of the corresponding structures are provided in the inset (from Ref. [34])

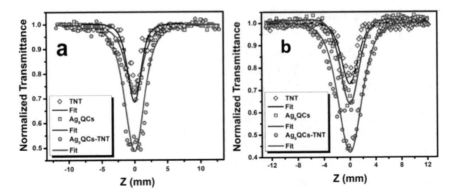

Fig. 6.7 Open aperture Z-scan curves measured for TNT, Ag9QCs and Ag9QCs-TNT. **a** Excitation by ultrashort (100 fs) laser pulses at 800 nm. Laser pulse energy used is 8 μJ. **b** Excitation by short (5 ns) laser pulses at 532 nm. Laser pulse energy used is 30 μJ (from Ref. [46])

have been used recently to design and fabricate nonlinear nanoantennas with controllable emission of the second harmonic (SH) radiation (Fig. 6.7) [34]. This work can be applied for the development of new types of nonlinear photonic components, which can be useful in integrated photonic circuits.

Applications of graphene in nonlinear optics continue to be explored vigorously. Second-order-nonlinear optical effects of graphene, particularly SHG, have been investigated theoretically [35–38]. Dean et al. have experimentally demonstrated SHG in graphene and multilayer graphite films mounted on a SiO_2/Si substrate using 150 fs laser pulses at 800 nm [39]. From numerical modeling, Gao et al. have demonstrated that phase matching and SHG can be realized in a pair of parallel graphene nanowires. The authors conclude that phase matching can be achieved between the hybridized modes by tuning the spacing of the nanowire pair. In order to achieve the largest conversion efficiency, one needs to choose the appropriate modes, and to further enhance the nonlinearity the nanowire radius should be decreased. Additionally, electromagnetic field enhancement due to surface plasmons will occur near the nanowires, further boosting the nonlinearity [40]. For frequency doubling applications, the abundance of polarizable π electrons in graphene could yield structures with large first hyperpolarizability if the centrosymmetry is broken. From computational modeling it has been shown that graphene nanoribbon (GNR) can be used as an excellent conjugated bridge in a donor–conjugated bridge–acceptor (D–B–A) framework to design high-performance second-order nonlinear optical materials [40].

6.7.2 Third Order Nonlinearities

Third-order nonlinear optical interactions (i.e., those described by a $\chi^{(3)}$ suscep-tibility) can occur for both centrosymmetric and non-centrosymmetric media. Symmetry-permitted third-order optical nonlinearity is remarkably strong in plas-monic, dielectric, carbon and graphene nanomaterials, leading to effects like saturable absorption, reverse saturable absorption and wave mixing. Third-order optical nonlin-earity plays a key role in nonlinear photonic devices. Materials which are currently relevant to third order nonlinear optics can be divided into five main categories, viz. metal nanoparticles, dielectric nanoparticles, carbon and graphene-based materials, and epsilon-near-zero (ENZ) materials.

6.7.2.1 Metal Nanoparticles

Regardless of the emission wavelength, ultrafast lasers utilize a mode-locking proce-dure, whereby a nonlinear optical component, called a saturable absorber, transforms the continuous-wave output into a train of ultrafast optical pulses [41, 42]. There are reports on the observation of saturable absorption (SA) in plasmonic metal nanoparti-cles, even though switching to reverse saturable absorption (RSA) has been observed at higher intensities [43, 44].

Noble metal nanostructures embedded in transparent dielectric matrices have been attracting significant attention towards all-optical signal processing device applica-tions. This is because of the large third-order nonlinear optical properties caused by the phenomena of surface plasmon resonance (SPR) and quantum size effect. Silver NPs experience a lower (compared to what?) intrinsic loss of plasmonic energy at visible frequencies. SA at 532 nm has been observed at low input irradiances in Ag nanodots prepared by pulsed laser deposition [45]. Kishore et al. [46] has studied enhanced nonlinear absorption of Ag_9QCs-TNT composite compared to its pristine precursors, Ag_9QC and TNT. Nonlinear absorption is attributed mainly to excited state and free carrier absorption phenomena for 5 ns excitation, while two-photon absorption should be more prominent for 100 fs excitation. α_2 measured for Ag_9QCs-TNT for ns excitation is 2.7×10^{-10} m/W, while that for fs excitation is 1.6×10^{-14} m/W. These values show that Ag_9QCs-TNT an excellent material for passive optical limiting device applications. Several studies are available in literature on Ag nanocomposites, some of which are shown in Table 6.1.

Gold metamaterials are the next promising candidates for NLO, which can be potentially used for fabricating optical sensors and saturable absorbers. In the pres-ence of external fields, they can control material properties as well [47]. Gold nanorods (GNRs) have been used as saturable absorbers for passive mode-locking at 1 μm wavelength. Absorption at the longitudinal surface plasmon resonance of GNRs is used to induce mode-locking. By using GNR film, stable passive mode-locking has been demonstrated at 1039 nm in an ytterbium-doped fiber laser cavity pumped by a 980 nm laser diode [48, 49]. Gold-nanospheres (GNS) also exhibit

Table 6.1 NLO investigations in some nanomaterials

Nanostructure	Wavelength (nm) ($\lambda_{\text{-probe}}$)/Pulse duration	Reference
Silver nanoparticles	532 nm, 6 ns	[93]
	1200 nm, 100 fs	[94]
	532 nm, 16 ns	[95]
	795 nm, 110 fs	[96]
	532 nm, 5 ns 800 nm, 100 fs	[97]
	532 nm, 5 ns	[98]
	532 nm, 5 ns	[99]
	800 nm, 190 fs	[100]
	800 nm, 100 fs	[101]
Gold nanoparticles	800 nm, 100 fs	[102]
	1500 nm, 70 fs	[103]
	532 nm, 26 ps	[104]
	800 nm, 150 fs	[105]
	532 nm, 8 ns	[106]
	532 nm, 35 ps	[76]
	800 nm, 35 fs	[107]
	532 nm, 5 ns	[44]
	532 nm, 7 ns	[108]
	800 nm, 60 fs	[109]
Dielectric materials Silicon composites	1200 nm, 240 fs	[110]
	1100–1500 nm, 100 fs	[111]
	532 nm, 8 ns	[112]
	772 nm, 275 fs 1030 nm, 140 fs 1550 nm, 97 fs	[113]
Two dimensional materials WS_2/MoS_2 composites	1030 nm, 340 fs 800 nm, 40 fs 515 nm, 340 fs	[114]
WS_2	800 nm, 100 fs	[115]
MoS_2	532 nm, 8 ns	[81]
Carbon/graphene based materials	400 nm–700 nm, 100 fs	[116]
	532 nm, 5 ns, 35 ps	[117]
	1053 nm, 75 ps	[118]
	400 nm, 800 nm, 1562 nm, 565 fs	[119]
	1030 nm, 340 fs	[120]
Epsilon zero materials	1240 nm, 150 fs	[121]
	720 nm, 200 fs	[122]

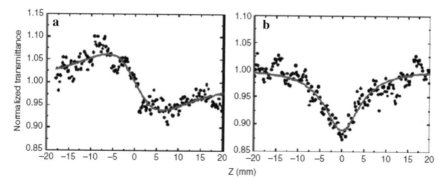

Fig. 6.8 Closed and open aperture Z-scan measurements of a gold metasurface at 1500 nm, using 70 fs, 1 kHz laser pulses obtained from an optical parametric amplifier. The peak intensity used is 45 GW/cm^2 (from Ref. [90])

saturable absorption behavior. An all-fiber passively Q-switched erbium doped fiber laser (EDFL) using GNS based saturable absorber (SA) with evanescent field interaction has been demonstrated [50]. Closed and open aperture Z-scan measurements carried out on a gold metasurface using 70 fs, 1500 nm laser pulses are shown in Fig. 6.8. From the data, the nonlinear optical parameters are calculated to be $\alpha_2 = 3 \times 10^{-6}$ cm/W and $n_2 = -1.05 \times 10^{-10}$ cm^2/W, respectively. Measurements carried out at 800 nm gave values of $\alpha_2 = -0.90 \times 10^{-4}$ cm/W and n2 $= -7.90 \times 10^{-9}$ cm^2/W. More detailed studies on gold nanocomposites are mentioned in Table 6.1 for further reference.

Other than Ag and Au nanoparticles, studies in some other promising metal nanoparticles such as Zn, Ni, and Al Ti, derivatives also are available in literature [51–54].

6.7.2.2 Dielectric Nanoparticles

Silicon-based optical devices have received much attention because of their potential applications in high-speed signal processing and no-chip communications [55]. A major advantage of silicon photonic devices is that they can be produced efficiently because of the highly advanced silicon processing technology, permitting low-cost, large-volume electronic circuit production. Silicon photonics makes optical devices compatible with CMOS technology, allowing for on-chip integration [56].

The nonlinear optical response of nc-Si/SiO$_2$ multilayers excited by ps and fs laser pulses, studied by Zhang et al. [57], is shown in Fig. 6.9. When the sample is excited by picosecond laser pulses absorption saturation is observed ($\alpha_2 = -3.1 \times 10^{-6}$ cm/W^{-1}), with a nonlinear refraction coefficient of $n_2 = -1.3 \times 10^{-10}$ cm^2 W^{-1}. Absorption saturation can be attributed to the single photon transition process between the valence band and the interface state, while negative nonlinear refraction is due to the free carrier dispersion effect. However, when the multilayers are excited

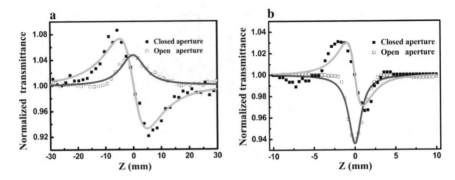

Fig. 6.9 Z-scan curves of nc-Si/SiO$_2$ multilayer sample in the closed aperture (full square) and open aperture (empty square) configurations. **a** $\lambda = 1064$ nm, $t_p = 25$ ps, laser intensity $I_0 = 1.18 \times 10^{10}$ W/cm^2. **b** $\lambda = 800$ nm, $t_p = 50$ fs, $I_0 = 3.54 \times 10^{11}$ W/cm^2. Solid curves are theoretical fits to the experimental data (from Ref. [91])

by femtosecond laser pulses, reverse saturation absorption is observed ($\alpha_2 = 1.1 \times 10^{-7}$ cm W^{-1}) while the sign of nonlinear refraction is unchanged ($n_2 = -1.5 \times 10^{-12}$ cm^2 W^{-1}). The difference in the free carrier densities generated by the different pulse durations play a significant role in the observed sign reversal of the absorptive nonlinearity. Variations in the nonlinear absorption and refraction responses of these multilayers during the transition from the amorphous to nanocrystalline phase also have been studied under femtosecond excitation at 800 nm by Zhang et al. [58]. These studies indicate the possible application of nc-Si/SiO$_2$ multilayers in photonic devices such as optical switches and Q-switch lasers.

6.7.2.3 Carbon and Graphene Based Materials

Carbon nanotubes and graphene have developed as promising materials for use in ultrafast fiber lasers [59, 60]. Their exceptional electrical and optical properties empower them to be utilized as saturable absorbers with fast responses and broadband operation, which can be effectively coordinated in fiber lasers. Assemblies of carbon nanotubes in suspension or in polymeric matrices can be used as saturable absorbers for near infrared light [61–63]. Currently used semiconductor saturable absorber mirrors have a thin tuning range, and require complex fabrication and packaging. A straightforward, cost-effective alternative is to utilize single-walled carbon nanotubes (SWNTs). Broadband tunability is achievable by utilizing SWNTs with a wide width distribution [64]. Since graphene–polymer composites are adaptable, they can be effortlessly incorporated into a number of photonic frameworks. So far, graphene–polymer composites [65, 66], CVD-grown films [67], functionalized graphene, and reduced graphene oxide flakes have been used for ultrafast lasers as saturable absorbers [68, 69]. Graphene-based ultrafast lasers and carbon-nanotube-based devices are discussed in the review paper by Bonaccorso et al. [70]. Martinez and Sun have demonstrated that nanotube and graphene nanocomposites are efficient saturable absorbers for fiber lasers. They emphasize that the easy fabrication and integration of SAs based on CNTs and graphene and their broad operation bandwidths are extremely valuable for various new fiber mode-locked lasers operating at broad wavelength ranges [71]. Saturable absorption has been shown in MoS_2, graphene and MoS_2/graphene nanocomposites in a number of solvents, as depicted in Fig. 6.10 [72].

Similarly, there is substantial enthusiasm for the study of optical limiting for protecting optical sensors and human eyes from intense light, as retinal harm can happen when intensities surpass a specific threshold. Passive optical limiters, which

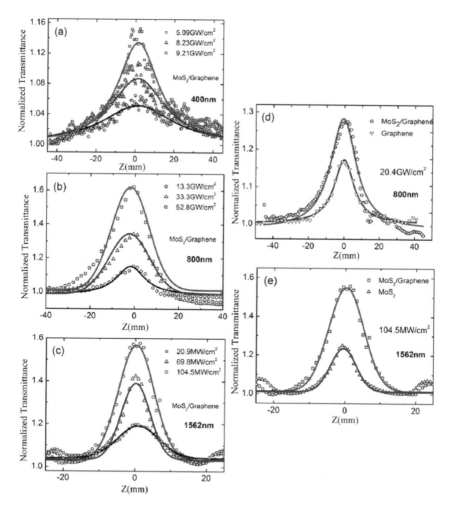

Fig. 6.10 Saturable absorption measured in MoS_2, graphene and MoS_2/graphene nanocomposites for different input fluences and wavelengths (from Ref. [72])

utilize nonlinear optical phenomena, can possibly be small, cheap and straightforward in design. However, no passive optical limiter has yet been designed which secures human eyes and common optical sensors from damage over the entire visible and near-infrared wavelength range. Reverse saturable absorption is a major physical mechanism which results in the optical limiting effect.

Sun et al. has reviewed optical limiters based on different classes of nanomaterials, including metal and semiconductor nanoparticles and nanoscale carbon materials [73]. They observe that carbon nanomaterials, ranging from carbon nanoparticles to fullerenes and to carbon nanotubes, are among the best optical limiters for nanosecond laser pulses. Philip and colleagues have reported that ligand-protected/capped gold and silver nanoclusters and their alloys are good optical limiters for nanosecond and picosecond laser pulses [44, 74–76]. Broad optical limiting in exfoliated graphene has been reported [45, 77, 78], and functionalized graphene dispersions are better than C_{60} for optical limiting [79, 80]. Carbon-based materials (for example, carbon-black dispersions, and CNTs, fullerenes and their derivatives) [81] have good optical limiting performance, in particular for nanosecond pulses at 532 and 1064 nm. The combination of plasmonic metal nanoparticles with carbon-based nanoparticles or different composites of graphene nanoparticles exhibit enhanced optical limiting [82–86] as seen from Fig. 6.11. Nonlinear absorption studies of different carbon-based materials/graphene composites are shown in Table 6.1.

6.7.2.4 Epsilon-Near-Zero (ENZ) Metamaterials

Recently, it has been found that materials with very small dielectric permittivity show efficient nonlinear optical behavior. These are generally known as epsilon-near-zero (ENZ) materials. These media can produce substantial enhancements to the local electric field, and high frequency conversion efficiencies [87, 88]. Transparent conducting oxides (TCO) have become important candidates in this category due to their unique NLO properties at epsilon-near-zero wavelengths. TCOs such as Indium-Tin-Oxide (ITO) exhibit a vanishing real part of the permittivity in the near-infrared where the group velocity of light decreases considerably and causes a strong light-matter interaction. A modest field intensity enhancement in the ITO film can lead to a large enhancement of nonlinear refraction coefficient (n_2) at the ENZ wavelengths. The hot-electron–induced optical nonlinearity of ITO films at ENZ wavelengths differs from that of noble metals under infrared irradiation in two ways. First, for a given change in permittivity, the nonlinear change in refractive index is always larger in the ENZ region than that in non-ENZ regions. Second, the free-electron heat capacity of ITO is more than one order of magnitude smaller than that of a noble metal such as gold. Thus, the increase in the free-electron temperature compared to the Fermi temperature and the consequent change in refractive index of ITO is much larger. ITO film shows an extremely large ultrafast third-order nonlinearity at ENZ wavelengths, and it can attain an optically induced change in the refractive index that is unprecedentedly large.

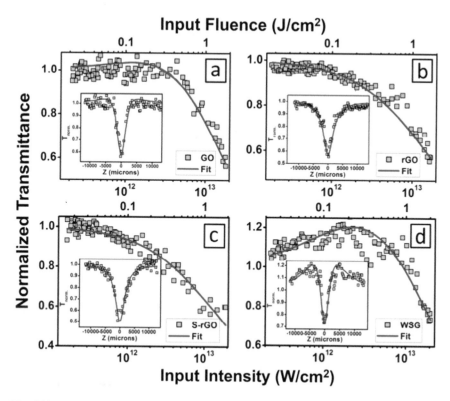

Fig. 6.11 Nonlinear transmission curves obtained from open aperture Z-scans (insets) measured for GO, rGO, S-rGO, and water soluble graphene (WSG), for 800 nm, 100 fs laser pulse excitation. Laser pulse energy used is 10 μJ (from Ref. [84])

ITO based sandwich structures with the insertion of silver (ITO/Ag/ITO) show large nonlinear optical enhancement of both nonlinear refraction and saturable absorption. The nonlinear refractive index ($n_2 = 15.43 \times 10^{-16}$ m^2/W) and nonlinear absorption coefficient ($\beta = -648 \times 10^{-11}$ m/W) measured for an ITO/Ag/ITO sandwich are found to be about 18 and 16 times greater than that measured for single-layer ITO, respectively, as shown in Fig. 6.12. The observed optical nonlinearity enhancement can be attributed to the inclusive effect of the increase of carrier concentration and local field enhancement induced by SPR in the sandwiches. Some of the investigations carried out with ITO hybrid metamaterials are shown in Table 6.1.

6.7.3 Applications of Optical Nonlinearity in Nanoparticles

There are different ways to classify NLO materials for different applications. The criteria can be the order of the nonlinearity, the nature of the NLO interaction, or some

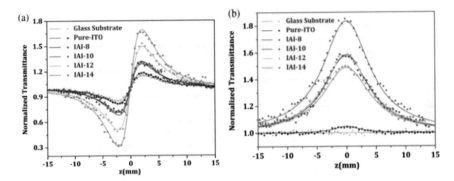

Fig. 6.12 Nonlinear optical responses of glass substrate, Pure-ITO, and ITO-Ag (IAI) sandwiches with various Ag insertion thicknesses, measured by Z-scan (from Ref. [92])

unique characteristics of the particular materials. Practically, for a given application, several potential NLO materials may be available.

Second-order NLO materials are used for the generation of new frequencies by means of SHG, sum- and difference-frequency mixing or optical parametric oscillation (OPO) [123]. In general, while several materials may be used for the production of radiation at a given wavelength, often one or two of them are better suited for a given process compared to the others. The second generation of NLO materials have been mainly organics, including organic single crystals, organic-inorganic single crystals, chromophore-hosted polymers and Langmuir-Blodgett (LB) films [124]. Third-order NLO materials comprise semiconductors, quantum confined semiconductors (quantum wells, quantum wires and quantum dots), metal particles, organics (crystals, polymers, and guest-host systems doped with active molecules) and inorganic glasses.

The development of different nanostructures has given a boost to nonlinear photonics, which has become an important research direction. Optical modulators, optical polarizers, optical switchers, and all-optical devices are part of nonlinear photonics research. The development of 2D layered materials was crucial in this regard, laying a good foundation for practical applications [125].

Another developing field of interest is nonlinear microscopy for biomedical applications using nanoparticles. Confocal fluorescence microscopy is a powerful tool for visualizing biological processes, but conventional laser scanning confocal microscopy cannot resolve structures below the diffraction limit of light. Recent advances in far-field super-resolution techniques have demonstrated that the diffraction limit can be overcome through a combination of smart optical design, novel photon excitation and detection schemes, the selection of specialized fluorescent labels, and dedicated data processing. Such nonlinear techniques can be sandwiched to the existing techniques to optimize the resolution of microscopy [126]. For instance, photon avalanche in lanthanide doped nanoparticles is useful for super-resolution imaging [127]. Developments are still going on in super-resolution microscopy to find better and faster methods for different applications [128]. Another promising line of work is optical nonlinear endoscopic tweezers, which is potentially an unprecedented tool for precisely specifying the location and dosage of drug particles, and for rapidly uploading metallic nanoparticles to individual cancer cells for treatment via two-photon absorption [129]. This technique allows an operation wavelength at the center of the transmission window of human tissue. Several other investigations also are ongoing on biomedical diagnostics and therapy using nonlinear optical responses of novel nanostructures [130, 131].

6.8 Conclusion

Large optical nonlinearities exhibited by dielectric, plasmonic and carbon-based nanomaterials hold great technological promise because of the possibility to modify their electronic and optical properties through material engineering. The emergence of metamaterials having novel structure-dependent properties has led to advanced research in nonlinear optics. Strong light-matter coupling in metamaterials produces novel linear and nonlinear optical properties. The development of these new materials in the field of photonics will continue to deepen, laying a good foundation for their practical applications. Selective combinations of dielectric, plasmonic and epsilon-near-zero materials is a very promising direction for further growth of nonlinear optics. These materials can be exploited for the fabrication of many key-devices for the telecom industry such as switches, routers, and wavelength converters. They are also attractive for applications such as photonic devices, sensing, terahertz imaging, and optical fibers. NLO nanomaterials can be used in the solution, film and solid forms, and a large number of chemical permutations and combinations are possible. In the medical field, nonlinear microscopy and nonlinear endoscopic tweezers hold great promise for diagnostics and treatment.

References

1. P.A. Franken, A.E. Hill, C.W. Peters, G. Weinreich, Phys. Rev. Lett. **7**, 118 (1961)
2. N. Bloembergen, Rev. Mod. Phys. **54**, 685 (1982)
3. R.W. Boyd, *Nonlinear Optics* (Academic Press, 2003)
4. P. Paufler, Cryst. Res. Technol. **26**, 802 (1991)
5. P.M. Rentzepis, Y. Pao, Appl. Phys. Lett. **5**, 156 (1964)
6. P.N. Prasad, *Nanophotonics* (Wiley, 2004)
7. B. Guo, Q.-L. Xiao, S.-H. Wang, H. Zhang, Laser Photonics Rev. **13**, 1800327 (2019)
8. A.K. Geim, K.S. Novoselov, Nat. Mater. **6**, 183 (2007)
9. J. Wang, Y. Chen, W.J. Blau, J. Mater. Chem. **19**, 7425 (2009)
10. R. Saito, G. Dresselhaus, M.S. Dresselhaus, Phys. Rev. B **61**, 2981 (2000)
11. G.Y.S.A.M. Nemilentsau, A.A. Khrushchinsky, S.A. Maksimenko, Carbon N. Y. **44**, 2246 (2006)
12. A.S.L. Gomes, M. Maldonado, L.S. de Menezes, L.H. Acioli, C.B. de Araújo, J. Dysart, D. Doyle, P. Johns, J. Naciri, N. Charipar, J. Fontana, Nanophotonics **9**(4), 725 (2020)

13. O. Reshef, I. De Leon, M. Zahirul Alam, R.W. Boyd, Nature Rev. Mater. **4**, 535 (2019)
14. M. Silveirinha, N. Engheta, Tunneling of electromagnetic energy through subwavelength channels and bends using ϵ-near-zero materials. Phys. Rev. Lett. **97**, 157403 (2006)
15. R.L. Sutherland, *Handbook of Nonlinear Optics* (CRC Press, 2003)
16. A. Hermans, C. Kieninger, K. Koskinen, A. Wickberg, E. Solano, J. Dendooven, M. Kauranen, S. Clemmen, M. Wegener, C. Koos, R. Baets, Sci. Rep. **7**, 44581 (2017)
17. S.K. Kurtz, T.T. Perry, J. Appl. Phys. **39**, 3798 (1968)
18. S. Chen, K.F. Li, G. Li, K.W. Cheah, S. Zhang, Light Sci. Appl. **8**, 17 (2019)
19. G. Berkovic, G. Meshulam, Z. Kotler, J. Chem. Phys. **112**, 3997 (2000)
20. K.Y. Suponitsky, S. Tafur, A.E. Masunov, J. Chem. Phys. **129.4**, 044109 (2008)
21. M. Sheik-Bahae, A.A. Said, T.-H. Wei, D.J. Hagan, E.W. Van Stryland, IEEE J. Quantum Electron. **26**, 760 (1990)
22. D.S. Correa, L.D. Boni, L. Misoguti, F.E.H.I. Cohanoschi, C.R. Mendonca, Opt. Commun. **277**, 440 (2007)
23. B. Gu, J. Wang, J. Chen, Y.-X. Fan, J. Ding, H.-T. Wang, Opt. Express **13**, 9230 (2005)
24. R.J. Gehr, R.W. Boyd, Chem. Mater. **8**, 1807 (1996)
25. P. Sheng, Phys. Rev. Lett. **45**, 60 (1980)
26. V.A. Markel, V.M. Shalaev, E.B. Stechel, W. Kim, R.L. Armstrong, Phys. Rev. B **53**, 2425 (1996)
27. M.V. Berry, I.C. Perciva, Opt. Acta **33**, 57728 (1986)
28. P.M. Hui, D. Stroud, Phys. Rev. **33**, 2163 (1986)
29. V.M. Shalaev, M.I. Stockman, Z. Phys. D: At. Mol. Clust. **10**, 71 (1988)
30. A.V. Butenko, V.M. Shalaev, M.I. Stockman, Zeitschrift Phys. D Atoms, Mol. Clust. **10**, 81 (1988)
31. M. Chandra, P.K. Das, Chem. Phys. **358**, 203 (2009)
32. M. Kauranen, A.V. Zayats, Nat. Photonics **6**, 737 (2012)
33. P.C. Ray, Chem. Rev. **110**, 5332 (2010)
34. G. Saerens, I. Tang, M.I. Petrov, K. Frizyuk, C. Renaut, F. Timpu, M. Reig Escalé, I. Shtrom, A. Bouravleuv, G. Cirlin, R. Grange, Laser Photonics Rev. **14**(9), 2000028 (2020)
35. J.J. Dean, H.M. van Driel, Phys. Rev. B **82**, 125411 (2010)
36. S.A. Mikhailov, Phys. Rev. B **84**, 45432 (2011)
37. M.M. Glazov, JETP Lett. **93**, 366 (2011)
38. S. Wu, L. Mao, A.M. Jones, W. Yao, C. Zhang, X. Xu, Nano Lett. **12**, 2032 (2012)
39. J.J. Dean, H.M. van Driel, Appl. Phys. Lett. **95**, 261910 (2009)
40. Y. Gao, I.V. Shadrivov, Sci. Rep. **6**, 38924 (2016)
41. Z.-J. Zhou, X.-P. Li, F. Ma, Z.-B. Liu, Z.-R. Li, X.-R. Huang, C.-C. Sun, Chem. A Eur. J. **17**, 2414 (2011)
42. Z. Kang, Q. Li, X.J. Gao, L. Zhang, Z.X. Jia, Y. Feng, G.S. Qin, W.P. Qin, Laser Phys. Lett. **11**, 35102 (2014)
43. Z. Kang, Y. Xu, L. Zhang, Z. Jia, L. Liu, D. Zhao, Y. Feng, G. Qin, W. Qin, Appl. Phys. Lett. **103**, 41105 (2013)
44. R. Philip, G.R. Kumar, N. Sandhyarani, T. Pradeep, Phys. Rev. B **62**, 13160 (2000)
45. R. Philip, P. Chantharasupawong, H. Qian, R. Jin, J. Thomas, Nano Lett. **12**, 4661 (2012)
46. U. Gurudas, E. Brooks, D.M. Bubb, S. Heiroth, T. Lippert, A. Wokaun, J. Appl. Phys. **104**(7), 073107 (2008)
47. K. Sridharan, P. Sankar, R. Philip, Opt. Mater. **94**, 53–57 (2019)
48. A.S. Gomes, M. Maldonado, L.D.S. Menezes, L.H. Acioli, C.B. de Araújo, J. Dysart, D. Doyle, P. Johns, J. Naciri, N. Charipar, J. Fontana, Nanophotonics **9**(4), 725–740 (2020)
49. Z. Kang, Q. Li, X.J. Gao, L. Zhang, Z.X. Jia, Y. Feng, G.S. Qin, W.P. Qin, Laser Phys. Lett. **11**(3), 035102 (2014)
50. J. Lee, J. Koo, J.H. Lee, Laser Phys. Lett. **14**(9), 090001 (2017)
51. D. Fan, C. Mou, X. Bai, S. Wang, N. Chen, X. Zeng, Opt. Express **22**(15), 18537–18542 (2014)
52. R. Sreeja, J. John, P.M. Aneesh, M.K. Jayaraj, Opt. Commun. **283**(14), 2908–2913 (2010)

53. E. Ramya, M.V. Rao, L. Jyothi, D.N. Rao, J. Nanosci. Nanotechnol. **18**(10), 7072–7077 (2018)
54. R. Kuladeep, L. Jyothi, P. Prakash, S. Mayank Shekhar, M. Durga Prasad, D. Narayana Rao, J. Appl. Phys. **114**(24), 243101 (2013)
55. R. Sato, S. Ishii, T. Nagao, M. Naito, Y. Takeda, ACS Photonics **5**(9), 3452–3458 (2018)
56. B. Jalali, Silicon photonics: nonlinear optics in the midinfrared, Nat. Photonics, **4**, 506–508 (2010)
57. D.A.B. Miller, Optical interconnects to electronic chips. Appl. Opt. **49**(25), 59–70 (2010)
58. P. Zhang, D.K. Li, L.Y. Jiang, J. Xu, vol. 8, 012012 (IOP Publishing, 2017)
59. P. Zhang, X. Zhang, J. Xu et al., Tunable nonlinear optical properties in nanocrystalline Si/SiO$_2$ multilayers under femtosecond excitation. Nanoscale Res. Lett. **9**(1), 28 (2014)
60. H.A. Haus, IEEE J. Sel. Top. Quantum Electron. **6**, 1173 (2000)
61. S. Tatsuura, M. Furuki, Y. Sato, I. Iwasa, M. Tian, H. Mitsu, Adv. Mater. **15**, 534 (2003)
62. G. Rozhin, Y. Sakakibara, H. Kataura, S. Matsuzaki, K. Ishida, Y. Achib, A.C Madoka Tokumoto, Chem. Phys. Lett. **405**, 288 (2005)
63. F. Wang, A.G. Rozhin, V. Scardaci, Z. Sun, F. Hennrich, I.H. White, W.I. Milne, A.C. Ferrari, Nat. Nanotechnol. **3**, 738 (2008)
64. T. Hasan, Z. Sun, F. Wang, F. Bonaccorso, P.H. Tan, A.G. Rozhin, A.C. Ferrari, Adv. Mater. **21**, 3874 (2009)
65. Z. Sun, T. Hasan, F. Torrisi, D. Popa, G. Privitera, F. Wang, F. Bonaccorso, D.M. Basko, A.C. Ferrari, ACS Nano **4**, 803 (2010)
66. H. Zhang, Q. Bao, D. Tang, L. Zhao, K. Loh, Appl. Phys. Lett. **95**, 141103 (2009)
67. H. Zhang, D.Y. Tang, L.M. Zhao, Q.L. Bao, K.P. Loh, Opt. Express **17**, 17630 (2009)
68. H. Zhang, D. Tang, R.J. Knize, L. Zhao, Q. Bao, K.P. Loh, Appl. Phys. Lett. **96**, 111112 (2010)
69. Y.-W. Song, S.-Y. Jang, W.-S. Han, M.-K. Bae, Appl. Phys. Lett. **96**, 51122 (2010)
70. W.D. Tan, C.Y. Su, R.J. Knize, G.Q. Xie, L.J. Li, D.Y. Tang, Appl. Phys. Lett. **96**, 31106 (2010)
71. F. Bonaccorso, Z. Sun, T. Hasan, A.C. Ferrari, Nat. Photonics **4**, 611 (2010)
72. A. Martinez, Z. Sun, Nat. Photonics **7**, 842 (2013)
73. Y. Jiang, L. Miao, G. Jiang et al., Broadband and enhanced nonlinear optical response of MoS2/graphene nanocomposites for ultrafast photonics applications. Sci. Rep. **5**, 16372 (2015)
74. Y.-P. Sun, J.E. Riggs, K.B. Henbest, R.B. Martin, J. Nonlinear Opt. Phys. Mater. **9**, 481 (2000)
75. S. Kumar, M. Anija, N. Kamaraju, K.S. Vasu, K.S. Subrahmanyam, A.K. Sood, C.N.R. Rao, Appl. Phys. Lett. **95**, 191911 (2009)
76. R.T. Tom, A.S. Nair, N. Singh, M. Aslam, C.L. Nagendra, R. Philip, K. Vijayamohanan, T. Pradeep, Langmuir **19**, 3439 (2003)
77. B. Karthikeyan, M. Anija, R. Philip, Appl. Phys. Lett. **88**, 53104 (2006)
78. R.P.M. Anija, J. Thomas, N. Singh, S. Nair, R.T. Tom, T. Pradeep, Chem. Phys. Lett. **380**, 223 (2003)
79. J. Wang, Y. Hernandez, M. Lotya, J.N. Coleman, W.J. Blau, Adv. Mater. **21**, 2430 (2009)
80. Y. Xu, Z. Liu, X. Zhang, Y. Wang, J. Tian, Y. Huang, Y. Ma, X. Zhang, Y. Chen, Adv. Mater. **21**, 1275 (2009)
81. L.W. Tutt, A. Kost, Nature **356**, 225 (1992)
82. N. Mackiewicz, T. Bark, B. Cao, J.A. Delaire, D. Riehl, W.L. Ling, S. Foillard, E. Doris, Fullerene-functionalized carbon nanotubes as improved optical limiting devices. Carbon **49**(12), 3998–4003 (2011)
83. S. Perumbilavil, P. Sankar, T. Priya Rose, R. Philip, Appl. Phys. Lett. **107**, 51104 (2015)
84. Z.-B. Liu, Y.-F. Xu, X.-Y. Zhang, X.-L. Zhang, Y.-S. Chen, J.-G. Tian, J. Phys. Chem. B **113**, 9681 (2009)
85. K. Sridharan, P. Sreekanth, T.J. Park, R. Philip, J. Phys. Chem. C **119**, 16314 (2015)
86. R.P. Sreekanth Perumbilavil, K. Sridharan, D. Koushik , P. Sankar, V.P. Mahadevan Pillai, Carbon N. Y. **111**, 283 (2017)

87. B. Anand, A. Kaniyoor, S.S.S. Sai, R. Philip, S. Ramaprabhu, J. Mater. Chem. C **1**, 2773 (2013)
88. B.S. Kalanoor, P.B. Bisht, S. Akbar Ali, T.T. Baby, S. Ramaprabhu, J. Opt. Soc. Am. B **29**, 669 (2012)
89. A. Ciattoni, C. Rizza, E. Palange, Extreme nonlinear electrodynamics in metamaterials with very small linear dielectric permittivity. Phys. Rev. A **81**, 043839 (2010)
90. A. Ciattoni, C. Rizza, E. Palange, Transmissivity directional hysteresis of a nonlinear metamaterial slab with very small linear permittivity. Opt. Lett. **35**, 2130–2132 (2010)
91. M. Kyoung, M. Lee, Opt. Commun. **171**, 145–148 (1999)
92. L.d.S. Menezes, L.H. Acioli, M. Maldonado, et al., Large third-order nonlinear susceptibility from a gold metasurface far off the plasmonic resonance, J. Opt. Soc. Am. B **36**, 1485 (2019)
93. P. Zhang, X. Zhang, L. Peng et al., Interface state-related linear and nonlinear optical properties of nanocrystalline Si/SiO$_2$ multilayers. Appl. Surf. Sci. **292**(1), 262–266 (2014)
94. K. Wu, et al., Large optical nonlinearity of ITO/Ag/ITO sandwiches based on Z-scan measurement. Opt. Lett. **44.10**, 2490–2493 (2019)
95. R. Sathyavathi, M.B. Krishna, S.V. Rao, R. Saritha, D.N. Rao, Adv. Sci. Lett. **3**(2), 138–143 (2010)
96. A. Alesenkov, J. Pilipavičius, A. Beganskienė, R. Sirutkaitis, V. Sirutkaitis, Lith. J. Phys. **55**(2) (2015)
97. Y. Deng, Y. Sun, P. Wang, D. Zhang, H. Ming, Q. Zhang, In situ synthesis and nonlinear optical properties of Ag nanocomposite polymer films. Phys. E. **40**(4), 911–914 (2008)
98. R.A. Ganeev, M. Baba, A.I. Ryasnyansky, M. Suzuki, H. Kuroda, Opt. Commun. **240**(4–6), 437–448 (2004)
99. K. Sridharan, T. Endo, S.G. Cho, J. Kim, T.J. Park, R. Philip, Opt. Mater. **35**(5), 860–867 (2013)
100. K.B. Bhavitha, A.K. Nair, S. Perumbilavil, S. Joseph, M.S. Kala, A. Saha, R.A. Narayanan, N. Hameed, S. Thomas, O.S. Oluwafemi, N. Kalarikkal, Opt. Mater. **73**, 695–705 (2017)
101. P.B. Anand, C.S. Sandeep, K. Sridharan, T.N. Narayanan, S. Thomas, R. Philip, M.R. Anantharaman, Adv. Sci. Eng. Med. **4**(1), 33–38 (2012)
102. K. Liu, X. Xu, W. Shan, D. Sun, C. Yao, W. Sun, Opt. Mater. **99**, 109569 (2020)
103. K. Yu, Y. Yang, J. Wang, X. Tang, Q.H. Xu, G.P. Wang, Nanotechnology **29**(25), 255703 (2018)
104. J. Fontana, M. Maldonado, N. Charipar, et al., Opt. Express **24**, 27360–27370 (2016)
105. L.D.S. Menezes, L.H. Acioli, M. Maldonado, J. Naciri, N. Charipar, J. Fontana, D. Rativa, C.B. de Araújo, A.S. Gomes, J. Opt. Soc. Am. B **36**(6), 1485–1491 (2019)
106. O. Sánchez-Dena, P. Mota-Santiago, L. Tamayo-Rivera, E.V. García-Ramírez, A. Crespo-Sosa, A. Oliver, J.A. Reyes-Esqueda, Opt. Mater. Express **4**(1), 92–100 (2014)
107. E.C. Romani, D. Vitoreti, P.M. Gouvêa, P.G. Caldas, R. Prioli, S. Paciornik, M. Fokine, A.M. Braga, A.S. Gomes, I.C. Carvalho, Opt. Express **20**(5), 5429–5439 (2012)
108. S. Qu, C. Du, Y. Song, Y. Wang, Y. Gao, S. Liu, Y. Li, D. Zhu, Chem. Phys. Lett. **356**(3–4), 403–408 (2002)
109. A. Rout, G.S. Boltaev, R.A. Ganeev, Y. Fu, S.K. Maurya, V.V. Kim, K.S. Rao, C. Guo, Nanomaterials **9**(2), 291 (2019)
110. S. Edappadikkunnummal, S.N. Nherakkayyil, V. Kuttippurath, D.M. Chalil, N.R. Desai, C. Keloth, J. Phys. Chem. C **121**(48), 26976–26986 (2017)
111. Y. Fu, R.A. Ganeev, P.S. Krishnendu, C. Zhou, K.S. Rao, C. Guo, Opt. Mater. Express **9**(3), 976–991 (2019)
112. A. Haché, M. Bourgeois, Appl. Phys. Lett. **77**(25), 4089–4091 (2000)
113. S. Minissale, S.E.L.Ç.U.K. Yerci, L. Dal Negro, Appl. Phys. Lett. **100**(2), 021109 (2012)
114. S. Vijayalakshmi, F. Shen, H. Grebel, Appl. Phys. Lett. **71**(23), 3332–3334 (1997)
115. S.R. Flom, G. Beadie, S.S. Bayya, B. Shaw, J.M. Auxier, Appl. Opt. **54**(31), F123–F128 (2015)
116. S. Zhang, N. Dong, N. McEvoy, M. O'Brien, S. Winters, N.C. Berner, C. Yim, Y. Li, X. Zhang, Z. Chen, L. Zhang, ACS Nano **9**(7), 7142–7150 (2015)

117. S. Mirershadi, F. Sattari, A. Alipour, S.Z. Mortazavi, Front. Phys. **8**, 96 (2020)
118. Z. Liu, Y. Wang, X. Zhang, Y. Xu, Y. Chen, J. Tian, Appl. Phys. Lett. **94**(2), 021902 (2009)
119. Z. Zheng, C. Zhao, S. Lu, Y. Chen, Y. Li, H. Zhang, S. Wen, Opt. Express **20**(21), 23201–23214 (2012)
120. Y. Jiang, L. Miao, G. Jiang, Y. Chen, X. Qi, X.F. Jiang, H. Zhang, S. Wen, Sci. Rep. **5**(1), 1–12 (2015)
121. P.L. Li, Y.H. Wang, M. Shang, L.F. Wu, X.X. Yu, Carbon **159**, 1–8 (2020)
122. M.Z. Alam, I. De Leon, R.W. Boyd, Science **352**, 795–797 (2016)
123. O. Reshef, E. Giese, M.Z. Alam, I. De Leon, J. Upham, R.W. Boyd, Opt. Lett. **42**(16), 3225–3228 (2017)
124. H.I. Elim, W. Ji, F. Zhu, Appl. Phys. B **82**, 439–442 (2006)
125. D.F. Eaton, Science **253**(5017), 281–287 (1991)
126. W. Nie, Adv. Mater. **5**(7–8), 520–545 (1993)
127. B. Guo, Q.L. Xiao, S.H. Wang, H. Zhang, Laser Photonics Rev. **13**(12), 1800327 (2019)
128. J. Squier, M. Müller, Rev. Sci. Instrum. **72**(7), 2855–2867 (2001)
129. A. Bednarkiewicz, E.M. Chan, A. Kotulska, L. Marciniak, K. Prorok, Nanoscale Horiz. **4**(4), 881–889 (2019)
130. B. Liu, C. Chen, X. Di, J. Liao, S. Wen, Q.P. Su, X. Shan, Z.-Q. Xu, L.A. Ju, C. Mi, F. Wang, D. Jin, Nano Lett. **20**(7), 4775–4781 (2020)
131. M. Gu, H. Bao, X. Gan, N. Stokes, J. Wu, Light: Sci. Appl. **3**(1), e126–e126 (2014)
132. M.E. Maldonado, A. Das, A.S.L. Gomes, A.A. Popov, S.M. Klimentov, A.V. Kabashin, Opt. Lett. **45**, 6695–6698 (2020)
133. Y. Liu, F. Wang, H. Lu, G. Fang, S. Wen, C. Chen, X. Shan, X. Xu, L. Zhang, M. Stenzel, D. Jin, Small **10**(1002), 201905572 (2020)

Conclusions

Four new technologies are going to rule the twenty-first century, out of which nanotechnology is the most promising one whose potentials are not yet fully unraveled. The miniaturization is the primary moto of the semiconductor industry while adding multi-functionality to the existing elements is the second major challenge. Better shelf life and threshold against the radiations are other special qualities which is essential for ensuring performing devices. Environmental toxicity is another major challenge of the semiconductor market, because most of the widely investigated, high performance semiconductor materials are proved to be severely toxic and hence scientists are in second thought about the use of mercury, Arsenic and Lead based materials in micro/nanoelectronics.

Keeping in mind the above mentioned challenges, scientists across the globe are trying to develop eco-friendly, cost effective, high shelf life nanomaterials with multifunctionality for functional electronic devices and other applications. Hence large scope exists in this area, in the coming decade.

Internet of Things (IoT) is going to invade our life sooner and the object to object communication is the key point in IoT. To make the non-smart devices into smart ones, the most important devices that are needed are sensors and transducers of all kinds. Piezoelectric sensors and optical sensors (including UV and IR sensors) are uniquely important in visualizing such an interconnected world.

Optical properties of nanomaterials were interesting topic of research in the last several decades because it serves as a probe for understanding the electronic band structure properties at the nanoscale and in confined dimensions. More recently, especially in the last decade, drastic revolutions happened in exploiting the special optical properties of the nanosctructured materials for the fabrication of functional devices.

Semiconductor nanomaterials are widely exploited for their applications in luminescent devices due to their strong fluorescence in the visible range. Novel nanomaterials and their core–shell combinations are being developed with high fluorescence quantum yield and multi-wavelength emission by playing with the particle size,

S. S. Nair and R. Philip, *Nanomaterials for Luminescent Devices, Sensors,
and Bio-imaging Applications*, Progress in Optical Science and Photonics 16,
https://doi.org/10.1007/978-981-16-5367-4

shell parameters and charges. Arsenic and cadmium free semiconducting materials are being tried for LED and laser applications and ZnS and their core–shell cousins emerge as the most promising material in this regard. Several spatial geometries are being developed like quantum cubes and cages in addition to the conventionally opted Quantum dots and quantum wires' geometry for fabricating functional devices. Development of an "all solution processed in-organic quantum dots based LEDs with no organic parts are the need of the hour which can revolutionize the LED displays and make the next generation QLED affordable to the common men's pocket. Numerous groups around the world including authors group are actively involved in the fabrication of such all solution processed 100% inorganic QLEDs.

Imaging is another exciting potential that semiconductor quantum dots can offer which is not just limited to the in vitro cellular imaging, but the scope is also widened to include IR mapping as well as live bio imaging for which novel cost-effective high shelf life, non-toxic semiconductor NPs are being developed with improved fluorescence quantum yield and threshold against photo-bleach. ZnS is a promising semiconducting material in this aspect considering its very low levels of toxicity and tunable multiwavelength emission covering near UV and Visible region. Authors group have demonstrated the imaging potential and high level of biocompatibility for ZnS based structures. Other materials like fluorescent carbon Nps and Graphene are also safe choices for the bio imaging applications.

Choice of materials for optical sensors and other related applications were most often tangled around the noble metals in the past few decades due to their strong visible SPR and colors and the multi log fold enhancement in Raman scattering intensity they can offer makes them often hot candidates for ultrasensitive optical detection of molecules. In the last few years, other novel economic alternatives to the noble metal NPs like gold, silver and platinum are being sought after for these applications. Copper emerges out as a very promising candidate as its SPR can be tuned towards the visible range by properly tailoring the size and shape of the particles. Visible range fluorescence and very low bio toxicity levels are also reported in copper nanostructures which enhances their application potential of them in bio imaging applications. Their characteristic SPR in IR region (exhibited by larger CuNps) makes it an ideal choice for live cancer IR mapping too. Authors group also have demonstrated the potential of Cu NPs in sensing, diagnostic and imaging applications. Shelf life and oxidation issues are the major challenges faced by researchers in employing these materials for sensing devices. Other materials like transition metals and their complexes are also being explored and to obtain visible range SPR is a great challenge.

Investigations on the special non linear optical properties of NPs and tapping of the exhibited special non linear properties for promising devices are the two major aspects which needs great attention. Two decades before, the special non linear optical properties of noble metals were reported for the first time which is followed by similar observations in fullerenes. Other NPs were also widely investigated for their optical non linearities and related applications. Reports on stronger optical non linearities and properties exhibited by NPs shows that the enhanced properties were mainly due to the large particle density, higher number of surfaces& charges and stability against agglomeration, as well as the modified band structure and density of states. Authors group is actively involved in research on these aspects and have proved the possibility of fabrication of active non linear optical devices like optical limiters and saturable absorbers, based on their observed properties. Shelf life, stability against agglomeration and gravitational settling, photo bleach threshold etc. are some points which needs special consideration at this point and novel materials are being tried with improved shelf life and stability against intense laser pulses.

Hence the "Optical Applications of Nanomaterials" is rather a very broad topic, covering whose entire aspects is out of scope for this book project. Authors have limited their discussions on the feasibility of employment of NPs for some sensing, luminescent & display devices and imaging applications. The last chapter is dedicated to the non linear optical applications of nanomaterials. Although authors tried to give a decent review of the R&D developments happened in the last few decades in these areas, some non deliberate omissions might have happened during citing some important works. Stability, shelf life and toxicity are the three major challenges in this area and development of novel materials with improved optical properties with out compromising these three essential pre requisites is the need of the hour.

Printed in the United States
by Baker & Taylor Publisher Services